CONTRACTOR'S GUIDE TO LEED CERTIFIED CONSTRUCTION

Gene Farmer, A.I.A., NCARB, LEED-AP BD+C

DELMAR
CENGAGE Learning

Australia • Brazil • Japan • Korea • Mexico • Singapore • Spain • United Kingdom • United States

Contractor's Guide to LEED Certified Construction

Gene Farmer, A.I.A., NCARB, LEED-AP BD&C

Vice President, Technology & Trades Professional Business Unit: Gregory L. Clayton

Executive Marketing Manager: Taryn Zlatin McKenzie

Acquisitions Editor: Bobby Person

Development: Dawn Jacobson

Editorial Assistant: Nobina Preston

Director of Marketing: Beth A. Lutz

Marketing Manager: Marissa Maiella

Marketing Coordinator: Rachael Torres

Senior Production Director: Wendy Troeger

Senior Content Project Manager: Betsy Hough

Senior Art Director: Benjamin Gleeksman

For product information and technology assistance, contact us at
Cengage Learning Customer & Sales Support, 1-800-354-9706

For permission to use material from this text or product, submit all requests online at **www.cengage.com/permissions.** Further permissions questions can be e-mailed to **permissionrequest@cengage.com**

Library of Congress Control Number: 2011935303

ISBN-13: 978-1-1110-3667-6

ISBN-10: 1-1110-3667-5

Delmar
5 Maxwell Drive
Clifton Park, NY 12065-2919
USA

Cengage Learning is a leading provider of customized learning solutions with office locations around the globe, including Singapore, the United Kingdom, Australia, Mexico, Brazil, and Japan. Locate your local office at: **international.cengage.com/region**

Cengage Learning products are represented in Canada by Nelson Education, Ltd.

To learn more about Delmar, visit **www.cengage.com/delmar**
Purchase any of our products at your local college store or at our preferred online store **www.cengagebrain.com**

NOTICE TO THE READER

Publisher does not warrant or guarantee any of the products described herein or perform any independent analysis in connection with any of the product information contained herein. Publisher does not assume, and expressly disclaims, any obligation to obtain and include information other than that provided to it by the manufacturer. The reader is expressly warned to consider and adopt all safety precautions that might be indicated by the activities described herein and to avoid all potential hazards. By following the instructions contained herein, the reader willingly assumes all risks in connection with such instructions. The publisher makes no representations or warranties of any kind, including but not limited to, the warranties of fitness for particular purpose or merchantability, nor are any such representations implied with respect to the material set forth herein, and the publisher takes no responsibility with respect to such material. The publisher shall not be liable for any special, consequential, or exemplary damages resulting, in whole or part, from the readers' use of, or reliance upon, this material.

Printed in the United States of America
1 2 3 4 5 6 7 15 14 13 12 11

*I would like to dedicate this book to my wife Audrey, who is my best friend,
soul mate, and inspiration for everything I do.*

*I would also like to thank Professor Ronald A. Baier for his reviews and assistance
in proofreading the manuscript and Jason Biondi for his help in
obtaining LEED documents.*

*I would like to especially thank Dawn Jacobson for her tireless efforts
in bringing this project to print.*

INTRODUCTION

Earth is our collective home and like our individual house it must be properly maintained. Our planet is currently suffering from the years of abuse and neglect it has been subjected to. We have polluted the air, the water, and the ground itself. The end result of this abuse is that our planet is sick. And as inhabitants of the earth we are becoming sick.

Apart from the airborne pollution spewed by industries in developed countries, one of the greatest problems facing the citizens in the United States is what to do with the waste generated each day. We live in a disposable society. Much of what we consume is packaged in material that eventually finds its way to the landfill. "The United States is sinking under a 'river of waste', says Lynn Landes, founder of Zero Waste America (ZWA).[1] In fact according to the EPA, there are over 254 million tons of nonhazardous waste generated in the United States each year.[2] Most of this waste finds its way to local and regional landfills. Some of these landfills are so large that they can be seen from space. It has been said the only two man-made objects that can be seen from space is the Great Wall of China and the Fresh Kills landfill in New York. Although EPA estimates indicate a decline in the number of active landfills, the sizes of the remaining active facilities are growing at an alarming rate.[3]

The waste deposited in these landfills include household trash and garbage, office waste including waste paper, cardboard, and office equipment, used tires, and construction debris. Construction and demolition (C&D) debris include waste from new construction activities, remodeling, and demolition or deconstruction. The waste stream from C&D operations accounts for approximately 160 tons of the waste per day. Of that, 60%–80% is being deposited in landfills.[4] There are currently no federal regulations regarding C&D waste unless it is considered hazardous, in which case the U.S. Environmental Protection Agency would regulate the methods of handling and disposal.

In 1998, the U.S. Green Building Council (USGBC) created the Leadership in Energy and Environmental Design (LEED) certification for buildings. The purpose of this program was to develop a national standard for energy efficient, healthy buildings. There are nine different LEED rating systems currently in use.[5] They are:

New Construction
LEED for New Construction and Major Renovations is designed to guide and distinguish high-performance commercial and institutional projects.

Existing Buildings: Operations & Maintenance
LEED for Existing Buildings: Operations & Maintenance provides a benchmark for building owners and operators to measure operations, improvements, and maintenance.

Commercial Interiors

LEED for Commercial Interiors is a benchmark for the tenant improvement market that gives the power to make sustainable choices to tenants and designers.

Core & Shell

LEED for Core & Shell aids designers, builders, developers, and new building owners in implementing sustainable design for new core and shell construction.

Schools

LEED for Schools recognizes the unique nature of the design and construction of K-12 schools and addresses the specific needs of school spaces.

Retail

LEED for Retail recognizes the unique nature of retail design and construction projects and addresses the specific needs of retail spaces.

Healthcare

LEED for Healthcare promotes sustainable planning, design, and construction for high-performance healthcare facilities.

Homes

LEED for Homes promotes the design and construction of high-performance green homes.

Neighborhood Development

LEED for Neighborhood Development integrates the principles of smart growth, urbanism, and green building into the first national program for neighborhood design.

Buildings meeting the LEED standards are awarded LEED certifications. There are currently over 4000 buildings that have already been awarded LEED certifications and a reported 24,000 in the certification pipeline.[5]

The certification consists of a four-tiered system ranging from LEED Certified to LEED Platinum. New buildings and major renovation projects certification is based on the building achieving a minimum of 26 credits to be certified, to a requirement of 56 credits to achieve the platinum level of certification. The credits, which have become the responsibility of both the design and construction teams, are in fact awarded in both the design and construction phases. Responsibility for compliance with the LEED requirements falls within the following four disciplines: Architecture, Civil Engineering, MEP Engineering, and Construction.

In June of 2009, the USGBC has launched a totally new version (LEED 3.0) of the requirements. The purpose of this book is to give the reader insight and guidance into how the construction professional can successfully operate within the LEED rating system. In addition, the author hopes that through a better understanding of the LEED process the constructor will overall develop a more sustainable construction practice.

PREFACE

The purpose of this book will be to educate construction professionals in the specific construction-related requirements they will be expected to be in compliance with and to propose strategies for undertaking their construction operations within the construction-related categories of LEED compliance.

This book is not intended to create converts to the world of sustainable construction by convincing the readers of advantages of sustainable building. It is understood that the readers of this book already have a basic knowledge of and an interest in this type of construction and that they are involved in or anticipating involvement in a LEED compliant project.

There are a multitude of books and other publications that address the subject of sustainable construction and LEED requirements. Many of these books provide a broad overview of the requirements for a LEED compliant building with the intention of educating the reader about LEED and the advantages of adopting stainable building philosophies. Others address the specific requirement of becoming a LEED-certified professional.

The current LEED standard for new construction consists of 57 credit point categories. Of these, only 33% are specific to the contractor's activities on the project. The other 66% of the point categories address requirements of the others including the design team. Although many books address the totality of LEED compliance, this book is intended to address the narrower scope of only the construction issues facing contractors working on LEED projects.

CONTENTS

Chapter 3: THE CONTRACTOR'S ROLE IN LEED PROJECTS 35

Chapter 4: SUSTAINABLE PROJECT MANAGEMENT 51

BACKGROUND AND ENVIRONMENTAL CONCERNS

DEVELOPMENT IN THE UNITED STATES

The United States of America is a little over 200 years old. During that time the country has faced many changes and challenges. A once totally natural ecosystem, the impact of man has made significant impacts on the once pristine environment. Our country's rapid increasing development, while small as a percentage of the total land mass, has created significant environmental issues. Development particularly over the past 50–75 years has been significantly driven by financial factors. Builders and developers have constructed as much as they could in as short a time as possible. The demand for housing, particularly single-family houses, in large suburban neighborhoods was in great demand during the 1950s, 60s and 70s. The flight of an upwardly mobile population from the urban areas to the suburbs presented an enormous profit making opportunity for builders and developers throughout the country.

Developers of that time gave virtually no consideration for how this rapidly increasing development might be permanently impacting the environment of our country. Coastal areas with their mangroves and estuaries became a prime focal point for the development of waterfront residential communities. Florida as well as other coastal states permanently lost large areas of coastal wildlife habitat during this time period. This lack of consideration has resulted in a country littered with thousands of inefficiently constructed and inefficiently operated urban sprawl developments with a significant loss of natural wildlife habitat.

According to the USDA Natural Resources Conservation Service, there are approximately 1.94 billion acres of land in the continental United States (see Figure 1-1).[1] This can be broken down as follows:

Figure 1-1

Land use in the United States.

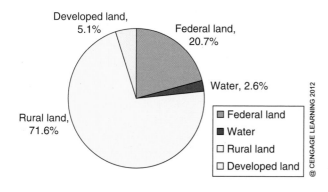

© CENGAGE LEARNING 2012

Federal land	402 million acres	20.7%
Water	50.4 million acres	2.6%
Rural land	1.377 billion acres	71.6%
Developed land	98 million acres	5.6%

Of the total 1.94 billion acres, the single largest area consists of privately owned rural land. This land is important because it provides support in the way of crops and food for the smaller developed areas. The land area that has been developed consists of approximately 98 million acres or 5.1% of the total land area. The problem facing the United States is the rapid expansion of the developed areas. Developed land in the United States is growing at a rate of 2.5% per year. This increase is approximately 2.65 times the increase in population.[2] Since the Federal land and the land that is covered with water cannot be generally developed, this expansion is occurring at the sacrifice of our country's rural areas. These rural areas, in addition to serving as a natural resource for farming and forestry, serve as a habitat for much of the wildlife in the United States. The damage caused to our country's wildlife habitats is creating significant pressures on the wildlife living there.

In addition to the large urban centers frequently found throughout the United States, our country is crisscrossed by a network of interstate highways. These highways effectively restrict both the natural flow of water across the surface of the land and the ability for animals intrinsic to the area to safely travel as necessary. There are currently 4 million linear miles of paved roads in the United States. This equals approximately 24,500 sq. mi. of paved surface. This is enough paving to cover one half of the sate of Pennsylvania. In addition to the paved roads, the United States has 10,900 sq. mi. of paved parking lots. Together the parking lots and the roads equal a paved surface of approximately 35,400 sq. mi. This equals the area of the state of Illinois. Like the previously discussed expansion of the nation's urban centers, the area used in our country for paving is also rapidly increasing. This increase is at a rate of approximately 3% per year, which again is endangering our country's natural habitats and wildlife (see Figures 1-2 and 1-3).[3]

Figure 1-2

Highways in the United States.

Figure 1-3

Growth of paved areas in the United States.

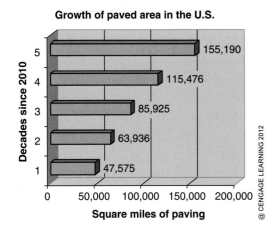

Growth of paved area in the U.S.

Decades since 2010 (vertical axis)

- 5: 155,190
- 4: 115,476
- 3: 85,925
- 2: 63,936
- 1: 47,575

0   50,000   100,000   150,000   200,000

Square miles of paving

@ CENGAGE LEARNING 2012

COMMERCIAL AND RESIDENTIAL CONSTRUCTION

According to a 2004 report issued by Energy Information Administration, there were approximately 4.9 million office buildings constructed in the United States as of 2003.[4] As cities expand, an increasing number of new buildings are needed. To fill this need, there are 170,000 new buildings constructed each year. Of these, 136,000 are new to the building inventory and the remaining are replacements for the 44,000 nonresidential buildings demolished annually.[5] Residential construction is also increasing, according to the American Housing Survey prepared in 2007 by the U.S. Department of Housing and Urban Development and the U.S. Department of Commerce; approximately 7.188 million new housing units were constructed between 2005 and 2009.[6] This puts the total number of housing units in the United States at over 130 million units. Of these, 63% or approximately 82 million units are single-family residences (see Figure 1-4).

In addition to the expanding commercial and business centers, these single-family residences also contribute significantly to the urban sprawl the United States has and is continuing to experience. According to the National Association of Home Builders, there are approximately 1.6 million new residences constructed each year. Of these, 1.355 million are totally new and the remaining are replacements for the 245,000 homes demolished annually. To support these new

Figure 1-4

Urban sprawl in the United States.

COURTESY OF SIMONP/WIKIPEDIA

residential neighborhoods, a vast number of new retail centers including open shopping centers and large covered malls have been constructed. These nonresidential support developments contribute greatly to the surface area impact of new development on the natural environment.

To complicate matters the sizes of our homes have been dramatically growing. From 1950 to 1999 the average size of a single-family residence in the United States has increased 105%. In 1950, the average size of a single-family residence was approximately 1110 sq. ft. In 1999, the average area had grown to 2250 sq. ft.[7] The impact of this increase in floor area is a significant increase in the materials required to construct these homes. Over 88% of the homes constructed in the United States are constructed of wood. It takes an average of 1.7 acres of forest to build one house on the United States. This is significant when compared with other parts of the world where a single-family residence can be constructed from as little as 0.7 acres of forest.[8] In addition to the additional material requirements these larger houses occupy more land area and consume considerably more energy to build and operate than their predecessors (see Figures 1-5 and 1-6).

Figure 1-5

Typical 1930s house.

COURTESY OF WWW.EMPSTUDIOS.NET

IMPACT ON OUR RESOURCES

There are several industries within the United States that historically have had a significant effect on the country's natural resources. One of the largest offenders is the mining industry that is responsible for the wholesale destruction of entire mountains. Even an industry such as the golf industry, which on the surface

Figure 1-6

Typical 2010 mansion.

Figure 1-7

Single-family residential construction systems.

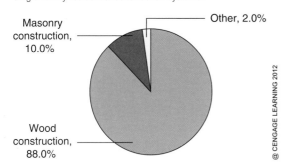

Masonry construction, 10.0%

Other, 2.0%

Wood construction, 88.0%

appears to be a rather benign industry, affects our natural resources. The volume of water pumped out of the ground to irrigate golf courses across the country is considered by some to be a practice that is unsustainable and potentially damaging.

The construction of the built environment in the United States, to a much lesser degree, also impacts three very important environmental areas, the country's natural resources, water, and air. The first impact is on our country's natural resources. The construction industry in the United States is one of the largest users of our country's natural resources. Approximately 88% of all homes constructed in the United States are constructed using wood-frame construction techniques. Ten percent employ the use of concrete masonry units in the exterior envelope. The remaining 2% of the residences are constructed of SIP panels, logs, or steel frame (see Figure 1-7).[9]

Although wood is a natural resource, most woods cannot be considered rapidly renewable, therefore conservation through recycling whenever possible is very important. Luckily wood is a relatively easy material to recycle. Recycled wood fiber is another wood-based product that is being increasingly used in construction. These wood fibers are used in materials ranging from structural beams and panels to interior finishes and cabinetry. Wood fiber is the primary component of oriented strand board (OSB), medium dense fiberboard (MDF), and particleboard. According to the Forest Products Journal, there are 29 million tons of wood fibers used in the United States each year.[10]

Water is another natural resource that is at peril. It is clear that the amount of water used in the actual construction process is insignificant and has very little impact on the general water supply. However, the actual construction process because of the disturbance to the natural ground patterns can have a significant effect on water resources. Construction sites are often great potential sources of soil erosion. Although not nearly on the scale of the damage caused by the mining industry, the construction process can have a detrimental effect on the water quality of a given geographic area. Once the natural pattern of the ground is changed and the inherent vegetation is removed, water runoff from the site and the deleterious effects of that runoff become an issue. This runoff is caused by rain or melting snow and can carry large amounts of soils into adjacent lakes, streams, or drainage structures. These soils and the chemicals they can often contain can have a significant detrimental effect on the wildlife in and around these bodies of water. Contractors must undertake significant precautions to mitigate the potential damages caused by this water runoff from their construction sites. This subject will be discussed in greater detail in Chapter 5.

The third natural resource impacted by construction is the outdoor air and atmosphere. It is true that the construction industry cannot be charged with the release of toxic chemical fumes like the chemical industry, steel industry, or paper industry. However, construction is a messy process that at times during high-wind conditions can be responsible for creating large clouds of dust. If left unchecked this dust can be a potential form of dangerous neighborhood pollution. In addition to localized pollution, many of the chemicals used directly in the construction process or indirectly in the materials used in our buildings can be hazardous to the environment. Some of these issues created by the construction process will be discussed later in this chapter and in Chapters 5 and 12.

WASTE GENERATION FROM CONSTRUCTION

In addition to the materials required to construct our country's residential and nonresidential buildings and their impact on the natural environment, construction operations generally result in the production of a significant amount of waste and debris. According to the Environmental Protection Agency (EPA) there are approximately 254 million tons of municipal solid wastes (MSWs) generated each year in the United States.[11] It is estimated that the total amount of construction waste generated by both construction and demolition (C&D) operations totals approximately 160 million tons annually.[12] That figure does not include waste resulting from road construction and paving operations, which if included would bring to a much higher-total. In all C&D waste accounts for approximately 26% of all nonindustrial waste production.

Construction Waste Recycling

The EPA estimates that the volume of MSW that is being recycled has consistently increased to its current volume of over 63.3 million tons. This indicates that approximately 25% of all MSW is currently being recycled. The EPA in its 2007 study also estimates that approximately 20%–30% of all C&D debris is currently being recycled.[13] This indicates that the recycling of C&D debris is keeping up with the recycling of other MSW. This subject will be discussed in greater detail in Chapter 9.

Construction Waste Breakdown

Figure 1-8

Construction and demolition waste source breakdown.

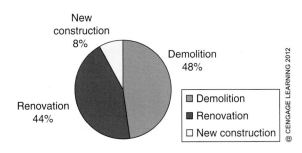

Construction waste is generated from two separate activities, construction and demolition, that is, C&D. Of the total 160 million tons of construction waste generated, it is estimated that 76.8 million tons or 48% is generated by demolition activities. Another 70.4 million tons or 44% are generated through renovation operations and the remaining 8% or 12.8 million tons are generated through new construction activities (see Figure 1-8).[14]

Broken down further, the construction of new buildings in the United States produces 2.5 lb. of waste per square foot of building area. This equals 250,000 lb. or 70,000 cu. ft. of waste on a typical 100,000 sq. ft. building.

Composition of Construction and Demolition Waste

The majority of all C&D waste falls into one of the following six categories:

1. Concrete and Masonry
2. Metals
3. Wood
4. Drywall
5. Roofing
6. Others

This includes cardboard, paper, interior finishes, carpeting and so on.

In 2007, DSM Environmental Services, Inc. was contracted by the Massachusetts Department of Environmental Protection (MassDEP) to undertake a study to identify the composition of C&D waste entering Massachusetts land fills. The results that cannot be directly attributed to land fill compositions in other states is relatively similar. The following chart illustrates the breakdown of C&D waste by weight (see Figure 1-9).

Wood is by far the largest component of C&D waste followed by roofing and drywall. The volume of concrete waste varies with the

Figure 1-9

Construction and demolition waste by weight.

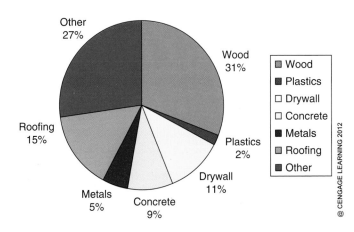

area of the country studied. Florida has a higher percentage of concrete waste than Washington due to the high percentage of concrete construction in Florida.

As will be discussed in Chapter 9, most of these materials can be readily recycled and this recycling will not only benefit the environment but there are also potential financial rewards available to the contractor who recycles a project's C&D waste.

TAKE NOTE

If you are not a part of the solution, you are a part of the problem. Be a problem solver.

Figure 1-10

Energy consumption by buildings in the United States.

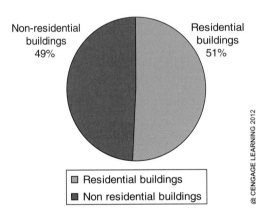

Non-residential buildings 49%

Residential buildings 51%

- Residential buildings
- Non residential buildings

@ CENGAGE LEARNING 2012

ENERGY CONSUMPTION

One of the single largest consumers of electrical energy in the United States is our country's buildings. According to the U.S. Department of Energy, in 2005 residential and nonresidential buildings consumed 38.9% of all the electricity produced in the United States. By 2006, this percentage had risen to 72% of all the electricity produced and it is expected to rise to 75% by the year 2025.[15] The EPA estimates that 51% of this electrical usage can be attributed to residential buildings, whereas 49% is used by nonresidential buildings (see Figure 1-10),

Residential Energy Consumption

According to the Energy Information Administration, heating and cooling our homes consume the largest portion of all residential energy consumption. The amount varies depending on the region but the percentage is considerable ranging from a low in the west of 43.8% to a high in the northeast of 62.4%. The next largest area of consumption is the heating of water. In almost every region, the energy consumed for water heating is slightly less than that consumed by all other appliance and lighting loads combined. The following chart illustrates the energy consumed for residential uses region by region across the United States (see Figure 1-11).

Commercial Energy Consumption

Energy consumption in the commercial sector is substantially greater than that found in the residential sector. This consumption varies widely with the building type. Hospitals are by far the largest consumers of electricity in the commercial

Figure 1-11

Average household energy usage.

2005 Delivered Energy End-Uses for an Average Household by Region (Million Btu per Household)					
	Northeast	Midwest	South	West	Average
Space Heating	71.8	58.4	21.0	26.3	40.5
Space Cooling	4.5	6.2	14.5	7.6	9.6
Water Heating	21.9	20.6	15.8	21.3	19.2
Refrigeration	4.3	4.9	4.8	4.3	4.6
Other Appliances and Lighting	23.0	25.9	25.0	24.1	24.7
TOTAL	122.2	113.5	79.8	77.4	94.9

U.S. ENERGY INFORMATION ADMINISTRATION, OCTOBER 2008

Source: EIA, A Look at Residential Energy Consumption in 2005, October 2008

sector. Hospitals consume approximately 250,000 btus/ft.^{2} of building area. This high energy consumption rate is understandable given the nature of the equipment used in hospitals and given the fact that they are operating 24 h. a day, 7 days a week. One energy consumption pattern that is clear is that like the residential sector the highest percentage of commercial building energy consumption is derived from three areas. These are space heating, cooling, and water heating. The following chart shows a breakdown by use category of five of the most common types of commercial facilities (see Figure 1-12).

It is clear that building designers and constructors must make a concerted effort to both design and construct buildings such that the amount of energy required to operate these buildings is reduced. Attention must be given to the building envelope as well as to the mechanical and electrical equipment selected for incorporation into the building.

TAKE NOTE

Our buildings will not become more energy efficient by themselves. Building designers and constructors must work together to improve the energy efficiency of our built environment.

Figure 1-12

Average commercial energy usage.

2003 Commercial Buildings Delivered Energy End Use Intensities (Thousands Btu per Square Foot)					
	Office	Retail	Education	Hospitals	Public assembly
Space Heating	32.8	24.8	39.4	91.8	49.7
Cooling	8.9	5.9	8.0	18.6	9.6
Ventilation	5.2	3.7	8.4	20.0	15.9
Water Heating	2.0	1.1	5.8	48.4	1.0
Lighting	23.1	25.7	11.5	40.1	7.0
Cooking	0.3	0.6	0.8	5.6	0.8
Refrigeration	2.9	5.0	1.6	2.0	2.2
Office Equipment	2.6	0.6	0.4	1.1	NA
Computers	9.0	5.6	4.0	18.1	6.5
Other	6.1	1.0	3.4	3.9	NA
TOTAL	92.9	73.9	83.1	249.2	93.9

2005 Delivered energy End-Use for as Average Household by Region (Million Btu per Household)					
	Northeast	Midwest	South	West	Average
Space Heating	71.8	58.4	21.0	26.3	40.5
Space Cooling	4.5	6.2	14.5	7.6	9.6
Water Heating	21.9	20.6	15.8	21.3	19.2
Refrigeration	4.3	4.9	4.8	4.3	4.6
Other Appliances and Lighting	23.0	25.9	25.0	24.1	24.7
TOTAL	122.2	113.5	79.8	77.4	94.9

U.S. ENERGY INFORMATION ADMINISTRATION, OCTOBER 2008

CONSTRUCTION IMPACT ON INDOOR AIR QUALITY

Unlike our ancestors, who often spent much of their time outdoors, Americans currently spend a majority of time indoors. During the mid-1990s, the EPA undertook a study of the relationship of the quality of indoor air to health. The study spanned across 10 EPA regions in 48 states. It was found that Americans spend as much as 87% of their time indoors. The study showed that of this time 69% of the time people were inside residences and 18% of the time they were inside other facilities like shops, schools, or offices.[16] Since most of our lives are spent inside, the quality

Figure 1-13

Average time spent in buildings.

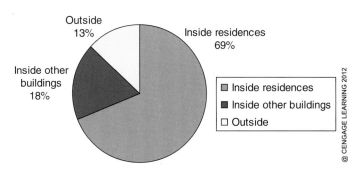

of the indoor environment has become increasingly more critical (see Figure 1-13).

Scientists are finding that except in our most polluted cities, the pollution levels of air inside our buildings can be as great as two to five times greater than that for the air outside the building. In rare cases, the pollution level inside a building has been measured as high as 100 times greater than that just outside the building. There are several reasons for this high level of indoor pollution, they include the following:

- Governmental requirement after the 1973 oil embargo to reduce the amount of outside air required to be brought into buildings.
- Improperly designed and installed air conditioning systems, resulting in an improper air distribution.
- Insufficient maintenance of the air conditioning systems sometimes resulting in chemical and biological contamination of the indoor air environment.
- Chemically saturated materials and finishes installed within the interiors of buildings resulting in long periods of off-gassing chemical vapors into the buildings air stream.

We now know that there is a direct correlation between the quality of air inside our buildings and the incidence of respiratory illness. According to the American Journal of Respiratory and Critical Care Medicine, there has been an alarming increase in respiratory illness in recent years. In fact, between the years 1980 and 1994, there was a 75% increase in cases of asthma in the United States. Currently, there are approximately 38% of all the people in the United States who suffer from asthma. In addition, studies are showing that the quality of indoor air not only affects health but also can affect worker performance. It has been estimated that there could be a savings of $23–56 billion in worker productivity achieved through increasing the air quality within our work spaces.

Legal Issues Involving Indoor Air

The current litigious society in which we live has seen an ever-increasing number of indoor-related cases reach the courts. The trend began with a 1990s landmark case in which several tenants of a building under renovation allegedly became sick and were unable to work. The interesting issue in this case was that the actual renovation was being undertaken on a different floor from where the tenants lived. The plaintiffs alleged that the renovation work contaminated the building's indoor environment causing them to become sick. The designers, engineers, contractors, and building operators were all subject to legal liability. The case was eventually settled for an undisclosed amount. The constructors might have been held liable for not taking sufficient efforts to prevent the construction contamination from spreading throughout the building.

In another case a Californian software company moved into a new office building. Immediately after moving in, several of the company's software engineers began complaining of headaches and respiratory problems. Like the previous example a renovation was being undertaken in another part of the building. A solvent-based adhesive was being used throughout the job. The fumes from this adhesive leaked into the building's ventilation system contaminating the software company's space. The software company eventually moved out and filed suit against the project's designers, the contractors, and the building owner. They eventually reached an out-of-court settlement of several million dollars.

Because cases such as these are becoming so common place, to avoid this type of legal liability, it is clearly very important that our buildings be designed and constructed of materials and systems that will not have a negative effect on the indoor air environment. Chapters 14–16 will discuss precautions that contractors must take during construction to limit the negative effects of the construction process on the building's indoor air environment. Chapter 12 will discuss the importance of selecting materials that will not have a detrimental effect on the indoor environment.

ENVIRONMENTAL IMPACT OF CONSTRUCTION

Just as a wool blanket keeps a persons' body warm at night, our planet is blanketed by its atmosphere. As certain gasses, referred to as "Greenhouse Gasses" are produced and released from our cities, they rise and are trapped inside the atmosphere. These trapped gasses create a dense blanket of gasses surrounding the earth. These "Greenhouse gasses" vary in their ability to trap heat. These gasses that envelope the earth like a large blanket consist of water vapor, carbon dioxide, nitrous oxide, ozone, methane, and chlorofluorocarbons. The primary three gasses are carbon dioxide, nitrous oxide, and methane. Not all greenhouse gasses are equal, for instance a molecule of nitrous oxide is able to contain 270 times more heat than a molecule of carbon dioxide. A molecule of methane can hold 21 times more heat than a molecule of carbon dioxide. The construction industry has been a major source of greenhouse gasses over the years. Freon gas used in almost all air conditioning systems until it was phased out starting from the mid-1990s. Greenhouse gasses, as they are called, are gasses which when emitted into the atmosphere deteriorate the earth's protective ozone layer. This protective layer serves to protect the earth from the harmful effects of sun's ultraviolet rays. Greenhouse gasses create holes in this layer, which in turn, allow the sun's harmful rays to reach the earth's surface. The following chart, which illustrates the types and amounts of greenhouse gasses in our atmosphere, shows that[17] CFC-11 and CFC-12 (Freon) have the greatest percentage increase in atmospheric concentrations since 1750 (see Figure 1-14).[18]

Carbon dioxide, the primary greenhouse gas, has increased approximately by 38% since the beginning of the industrial revolution in 1750. Some scientists are predicting another increase of up to 30% within the next 50 years. Commercial buildings and residences in North America, the United States, and Canada contribute approximately 2.2 billion tons of carbon dioxide into the atmosphere

Figure 1-14

History of greenhouse gas levels in the atmosphere.

Gas	Pre 1750 Troposphereic Concentrations	Recent Troposphereic Concentrations
Concentration in Parts per Million ppm		
Carbon dioxide (CO_2)	280	384.8
Concentration in Parts per Billion ppb		
Methane (CH$_4$)	700	1865
Nitrous oxide (N$_2$O)	270	322
Troposphereic ozone (O$_3$)	25	34
Concentration in Parts per Trillion ppt		
CFC-11 (freon, used in air conditioning systems)	0	244
CFC-12 (refrigerant and aerosol spray propellant)	0	0538
Halon-1211 (used in fire extinguishers)	0	4.4
Halon-1301 (used in building fire suppression systems)	0	3.3
Carbon tetrachloride (a cleaning agent)	0	89

© CENGAGE LEARNING 2012

each year.[19] The trend is continuing according to the EPA, greenhouse gas emissions in the United States have increased 17% from 1990 to 2007.[20] It is imperative that we design and construct our buildings using materials and methods that will limit their impact on the environment. This book is intended to serve as a guide to help those who construct our buildings, with a minimum of environmental impact.

INTRODUCTION TO LEADERSHIP IN ENERGY AND ENVIRONMENTAL DESIGN

BACKGROUND

U.S. Green Building Council (USGBC), a private, nonprofit corporation, was founded in 1993 by David Gottfried, with the goal of promoting more environmentally friendly design and construction in the United States. Though it uses the initials U.S., it is not an agency of the government of the United States. From its inception, USGBC has grown steadily to its current community that is comprised of 78 local affiliates and 18,000 member companies and organizations. In addition to its members, USGBC has over 140,000 Leadership in Energy and Environmental Design (LEED) professional credential holders. To achieve this status, these credential holders have received education in the fundamentals of the LEED rating system and have successfully completed a stringent examination.

Shortly after its formation, USGBC began the development of the LEED. The LEED rating system is a procedure for evaluating a sustainable or "green" building's design, construction, and performance.

According to USGBC, LEED was developed with the following six primary goals in mind:

1. Define "green building" by establishing a common standard of measurement.
2. Promote integrated, whole-building design practices.
3. Recognize environmental leadership in the building industry.
4. Stimulate green competition.
5. Raise consumer awareness of green building benefits.
6. Transform the building market.

Currently there are over 35,000 projects across the United States and 91 foreign countries, which are participating in the LEED program. These projects comprise over 4.5 billion sq. ft. of building floor area.

TAKE NOTE

The value of green building construction is estimated to be $60 billion.
The green building market is estimated to be worth $30–$40 billion annually.

LEED Version 1.0 & 2.0

In August 1998, USGBC launched the first LEED version "LEED NCv1.0." This first pilot version for new construction "NC" underwent 2 years of testing and review. This review leads to significant modifications. In March 2000, the LEED Green Building Rating System for New Construction and Major Renovations, Version 2.0 (LEED NCv2.0), was released. This new version evaluates a building's performance from a whole building perspective. In 2005, LEED NCv2.2 was released. Currently LEED 2009 consists of nine rating systems organized into five categories.

INTRODUCTION TO LEED 2009 V 3.0

On April 27, 2009, USGBC released LEED 2009 v 3.0. This newly revamped system is considerably different from its predecessor. The new standard has nine different rating systems organized into five categories. They are as follows:

LEED BUILDING DESIGN AND CONSTRUCTION

- LEED for New Construction and Major Renovations: The intent of this rating system is similar to the previous version in that it applies to the use of sustainable techniques in the design and construction of new buildings, as well as those which are undergoing major renovations.
- LEED for Core and Shell: Like the previous version, this rating system is employed when the proposed project involves the use of sustainable design and construction techniques applicable to the building's core and exterior shell.
- LEED for Schools: This rating system applies to the use of specific sustainable techniques in the design and construction of K-12 educational facilities.
- LEED for Retail New Construction: This rating system is similar to LEED NC but it applies to exclusively retail projects. It is scheduled for release in 2010.

LEED INTERIOR DESIGN AND CONSTRUCTION

- LEED for Commercial Interiors: Like the previous version, this rating system is used when the proposed project involves the interior modification of a commercial building, such as in tenant improvements.

- LEED for Retail Interiors: This rating system is similar to the commercial interiors version except that it addresses the construction of retail interiors. It is scheduled for release in 2010.

LEED BUILDING OPERATIONS AND MAINTENANCE

- LEED for Existing Buildings Operations and Maintenance: This rating system applies to the continued operations and maintenance of existing buildings.

LEED NEIGHBORHOOD DEVELOPMENT

- LEED for Neighborhood Development rating system: This rating system is used to certify neighborhood developments that incorporate the principles of sustainability and smart growth.

LEED HOME DESIGN AND CONSTRUCTION

- LEED for Homes: This rating system is used in the certification of high performance residential projects.

LEED 2009 v3.0 Credits

The number of credit categories in LEED 2009 v3.0 has been increased from six to seven. In addition the credits are weighted according to regional priorities (RP). The credit categories of LEED 2009 v3.0 are as follows:

Sustainable Sites (26 points)

Sustainable Sites.

Choosing a building's site and managing that site during construction are important considerations for a project's sustainability. This credit category discourages the development in previously undeveloped land; minimizes a building's impact on ecosystems and waterways; encourages regionally appropriate landscaping; rewards smart transportation choices; controls stormwater runoff; and reduces erosion, light pollution, heat island effect, and construction-related pollution.

© CENGAGE LEARNING 2012

Water Efficiency (10 points)

Buildings are major users of our potable water supply. The goal of this credit category is to encourage smarter use of water, inside and out. Water reduction is typically achieved through more efficient appliances, fixtures and fittings inside, and water-wise landscaping outside.

Water Efficiency.

Energy and Atmosphere (35 points)

According to the U.S. Department of Energy, buildings use 39% of the energy and 74% of the electricity produced each year in the United States. This category encourages a wide variety of energy strategies: commissioning; energy use monitoring; efficient design and construction; efficient appliances, systems and lighting; the use of renewable and clean sources of energy, generated on-site or off-site; and other innovative strategies.

Energy and Atmosphere.

Materials and Resources (14 points)

During both the construction and operations phases, buildings generate a lot of waste and use a lot of materials and resources. This credit category encourages the selection of sustainably grown, harvested, produced, and transported products and materials. It promotes the reduction of waste as well as reuse and recycling, and it takes into account the reduction of waste at a product's source.

Materials and Resources.

Indoor Environmental Quality (15 points)

The U.S. Environmental Protection Agency estimates that Americans spend about 90% of their day indoors, where the air quality can be significantly worse than outside. This credit category promotes strategies that can improve indoor air as well as providing access to natural daylight and views and improving acoustics.

Indoor Environmental Quality.

Innovation in Design

Innovation in Design.

This credit category provides bonus points for projects that use new and innovative technologies and strategies to improve a building's performance well beyond what is required by other LEED credits or in green building considerations that are not specifically addressed elsewhere in LEED. It also rewards projects for including a LEED Accredited Professional on the team to ensure a holistic, integrated approach to the design and construction phase. There are six additional credits available in this category.

Regional Priority

Regional Priority.

USGBC's regional councils, chapters, and affiliates have identified the environmental concerns that are locally most important for every region of the country, and six LEED credits that address those local priorities were selected for each region. A project that earns a RP credit will earn one bonus point in addition to any points awarded for that credit. Up to four extra points can be earned in this way. See the RP Credits for your state

What LEED Measures, USGBC.org (All the images above Courtesy of U.S. Green Building Council—LEED Reference Guide)

Regional Priorities

Urban Florida.

Rural Michigan.

RP is a new category added in this version. This new category takes into consideration the environmental priorities of different regions within the United States. An illustration of these RPs is found in the LEED 2009 Reference book and is shown below:

Examples from LEED for New Construction 2009: (2.1)

Urban Florida: SSc5.2, MRc1.1, WEc2, EAc1, MRc5, and EQc8.1, to incentivize (among other things) decreased reliance on fossil fuels, reuse of existing building stock, decreased reliance on insufficient municipal wastewater plants, and utilization of abundant local sunshine.

Rural Michigan: SSc1, SSc6.1, SSc6.2, SSc8, MRc5.2, and EAc2, to incentivize (among other things) the preservation of prime agricultural land, reduction of light trespass into neighboring natural habitats, and minimizing the amount and improving the quality of storm water into the Great Lakes.

Priorities for each state are listed on the USGBC Web site at: http://www.usgbc.org

LEED Credit Weightings

USGBC states that, "Another major advancement that comes with LEED 2009 v3.0 is that credits will now have different weightings depending on their ability to impact different environmental and human health concerns. With revised credit weightings, LEED now awards more points for strategies that will have greater positive impacts on what matters most—energy efficiency and CO_2 reductions. Each credit was evaluated against a list of 13 environmental impact categories, including climate change, indoor environmental quality, resource depletion, and water intake, among many others. The impact categories were prioritized, and credits were assigned a value based on how they contributed to mitigating each impact. The result revealed each credit's portion of the big picture, giving the most value to credits that have the highest potential for making the biggest change. The credits are all intact; they are just worth different amounts."[1]

The following taken from USGBC better defines the weighting equation (see Figue 2-1).[2]

Figure 2-1

Basic weight equation.

Relative importance of each impact category

x

Relative contribution of a building **activity group** to building impacts

x

Association between individual credits and activity groups

= Credit Weight

COURTESY OF U.S. GREEN BUILDING COUNCIL—LEED REFERENCE GUIDE

Definitions

- **Impact category:** impacts of building on environment and occupants (e.g., TRACI categories).
- **Activity group:** a building-related function associated with a group of LEED credits (e.g., consumption of energy by building systems, transportation, and water use).
- **Association** with activity group: a binary (yes/no) relationship indicating whether or not a credit contributes to reducing an impact.

The categories and the weights assigned include the following:

- Greenhouse gas emissions (29%)
- Eutrophication (6%)
- Human health cancer (8%)
- Fossil fuel depletion (10%)
- Ozone formation (2%)
- Human health noncancer (5%)
- Water use (8%)
- Smog formation (4%)
- Indoor air quality (3%)
- Land use (6%)
- Ecotoxicity (7%)
- Acidification (3%)
- Particulates (9%)

The strategies used on any given project will be determined by the designers in consultation with the other members of the design and construction team.

LEED Certification

Finally, the USGBC has revised the number of credits required to achieve the various levels of certification. The revised credit requirements are as follows:

- LEED Certified 40–49 points
- LEED Silver 50–59 points
- LEED Gold 60–79 points
- LEED Platinum 80 points and above

The credits required to achieve the lowest level of certification, "LEED Certified," has increased from 26 points in the previous version to 40 points in this new version. The point threshold to achieve the highest certification, Platinum, has increased from 52 points in the previous version to a level of 80 points in the current version.

LEED VERSION 3.0 REQUIREMENTS

The requirements for LEED certification, as mentioned earlier, are broken down into seven categories. There are two sublevels of requirements within each category. The first sublevel of requirement is the "Prerequisite." Prerequisites are mandatory requirements that must be achieved by all projects. There are no

Figure 2-2

Prerequisites.

Credit	Description	Points
Sustainable Sites		
Pre-1	Construction Activity Pollution Prevention	Req'd
Water Efficiency		
Pre-1	Water Use Reduction- 20% Reduction	Req'd
Energy and Atmosphere		
Pre-1	Fundamental Commissioning of Building Energy Systems	Req'd
Pre-2	Minimum Energy Performance	Req'd
Pre-3	Fundamental Refrigerant Management	Req'd
Materials and Resources		
Pre-1	Storage and Collection of Recyclables	Req'd
Indoor Environmental Quality		
Pre-1	Minimum Indoor Air Quality Performance	Req'd
Pre-2	Environmental Tobacco Smoke (ETS) Control	Req'd

credits awarded, so compliance will not advance the project numerically toward certification. Within the seven categories, there are 10 prerequisites. The following figure lists these prerequisites (see Figure 2-2).

All categories except "Innovation in Design and Regional Priority" have at least one prerequisite. The second level of requirements is the credits. Each category has between 4 and 19 credits, which can be achieved through compliance with the standard. In all there are 60 different credit-awarding categories, which any project can pursue. The Owner, in consultation with the project team, will decide which of these credit categories the project will aspire to obtain. As more than one credit is achievable in some categories, ultimately, there are 100 possible points achievable, 110 when Innovation in Design and Regional Priority points are included. They are as follows (see Figure 2-3).

Credit	Description	Points
Sustainable Sites (26 possible points)		
Pre-1	Construction Activity Pollution Prevention	Req'd
SS-1	Site Selection	1
SS-2	Development Density and Community Connectivity	5
SS-3	Brownfield Redevelopment	1
SS-4.1	Alternative Transportation: Public Transportation Access	6
SS-4.2	Alternative Transportation: Bicycle Storage and Changing Rooms	1
SS-4.3	Alternative Transportation: Low-Emitting and Fuel-Efficient Vehicles	3
SS-4.4	Alternative Transportation: Parking Capacity	2
SS-5.1	Site Development: Protect or Restore Habitat	1
SS-5.2	Site Development: Maximize Open Space	1
SS-6.1	Stormwater Design: Quantity Control	1
SS-6.2	Stormwater Design: Quality Control	1
SS-7.1	Heat Island Effect: Non-Roof	1
SS-7.2	Heat Island Effect: Roof	1
SS-8	Light Pollution Reduction	1
Water Efficiency (10 possible points)		
Pre-1	Water Use Reduction	Req'd
WE-1	Water-Efficient Landscaping	2–4
WE-2	Innovative Wastewater Technologies	2
WE-3	Water Use Reduction	2–4
Energy and Atmosphere (35 possible points)		
Pre-1	Fundamental Commissioning of Building Energy Systems	Req'd
Pre-2	Minimum Energy Performance	Req'd
Pre-3	Fundamental Refrigerant Management	Req'd
EA-1	Optimize Energy Performance	1–19
EA-2	On-Site Renewable Energy	1–7
EA-3	Enhanced Commissioning	2
EA-4	Enhanced Refrigerant Management	2
EA-5	Measurement and Verification	3
EA-6	Green Power	2

Figure 2-3 *(Continues)*
LEED 2009 v 3.0 credits.

Materials and Resources (14 possible points)		
Pre-1	Storage and Collection of Recyclables	Req'd
MR-1/1.1	Building Reuse-Maintain Existing Walls, Floors, and Roof	1 to 3
MR-1.2	Building Reuse-Maintain 50% Interior Non-structural Elements	1
MR-2	Construction Waste Management	1–2
MR-3	Materials Reuse	1–2
MR-4	Recycled Content	1–2
MR-5	Regional Materials	1–2
MR-6	Rapidly Renewable Materials	1
MR-6/7	Certified Wood	1
Indoor Environmental Quality (15 possible points)		
Pre-1	Minimum Indoor Air Quality Performance	Req'd
Pre-2	Environmental Tobacco Smoke (ETS) Control	Req'd
IEQ-1	Outdoor Air Delivery Monitoring	1
IEQ-2	Increased Ventilation	1
IEQ-3/3.1	Construction Indoor Air Quality Management Plan-During Construction	1
IEQ-3.2	Construction Indoor Air Quality Management Plan-Before Occupancy	1
IEQ-4.1	Low-Emitting Materials-Adhesives and Sealants	1
IEQ-4.2	Low-Emitting Materials-Paints and Coatings	1
IEQ-4.3	Low-Emitting Materials-Flooring Systems	1
IEQ-4.4	Low-Emitting Materials-Composite Wood and Agrifiber Products	1
IEQ-5	Indoor Chemical and Pollutant Source Control	1
IEQ-6.1	Controllability of Systems-Lighting	1
IEQ-6/6.2	Controllability of Systems-Thermal Comfort	1
IEQ-7/7.1	Thermal Comfort-Design	1
IEQ-7.2	Thermal Comfort-Verification	1
IEQ-8.1	Daylight and Views-Daylight	1
IEQ-8.2	Daylight and Views-Views	1
Innovation In Design (6 possible points)		
ID-1	Innovation In Design	1–5
ID-2	LEED ® Accredited Professional	1
Regional Priority (4 possible points)		
RP-1	Regional Priority	1–4

Figure 2-3 *(Continued)*
LEED 2009 v 3.0 credits.

LEED PROJECT CERTIFICATION PROCESS

The initial decision that must be made in the certification process is the responsibility of the owner. The owner must decide which particular LEED rating system will be most applicable to their proposed project. The Green Building Certification Institute (GBCI) Web site, located at http://www.gbci.org, can provide guidance as to which rating system will be most appropriate for the project. Once the rating system has been selected, the path to certification is a five step process (see Figure 2-4).

Figure 2-4

LEED certification process.

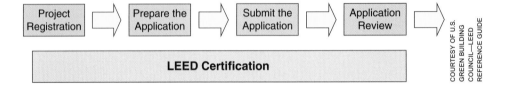

Step One: Project Registration

The first step of the LEED certification process involves the registration of the project with the GBCI. The GBCI has assumed the administrative responsibility for all LEED project certifications. Project registration must be completed online through the LEED online Web site at www.leedonline.com. In fact LEED online is the primary source for information on LEED requirements. It serves as a clearinghouse through which the LEED certification process for a registered project is managed. Through LEED online the certification team can do the following:

- Submit documentation to USGBC for review.
- Document compliance with LEED Credit Requirements.
- Coordinate resources among project team members.
- Manage public facing project details.
- Submit technical inquiries regarding LEED Credits.
- Track progress toward LEED Certification.

In order to register a project, the registrants must first register themselves as a site user. This is a simple process that can be accomplished in about 3 min. at no cost. The project registrant becomes the default project administrator. This person will have the responsibility of submitting all required information throughout the certification process by himself or herself or this person can delegate this responsibility to the design and construction team. To do this all persons responsibly for submitting information on the project must also register as such, online.

The registration of a project is a five step process, which includes the following:

- Eligibility
  This involves verifying that the project meets all of the requirements to be considered for LEED Certification.

- Rating System Selection
  This step involves the actual selection of the rating systems to be used for certification.
- Rating System Results
  In this step, the project administrator receives an online LEED "Scorecard" that identifies all of the prerequisites and possible credits in the selected rating system. See Appendix-A for a sample of this scorecard.
- Project Information
  In this step, the project administrator inputs information on the project itself. The input items include the following:
  1. Project title
  2. Project address
  3. Anticipated construction start date
  4. Anticipated construction finish date
  5. Rating system selected
  6. Gross Square Footage
  7. Anticipated project type
  8. Anticipated certification level
- Review
  In this step, the computer reviews the project information and registration information for completeness. If complete, the computer requests a specific registration fee.
- Payment
  This next to last step involves the project administrator paying the registration fee (see Figure 2.1 for a schedule of registration fees).
- Confirmation
  In this last step, the computer issues a final conformation that the project is registered, and the project is issued a registration number (see Figure 2-5).

	Less Than 50,000 sq. ft.	50,000–500,000 sq. ft.	More Than 500,000 sq. ft.	sq. ft. Appeals (If Applicable)
LEED 2009; New Construction, Commercial Interiors, Schools, Core & Shell full certification	**Fixed Rate**	**Based on Square Footage**	**Fixed Rate**	**Per Credit**
Design Review				
USGBC Members	$2000	$0.04/sf	$20,000	$500
Non-Members	$2250	$0.045/sf	$22,500	$500
Expedited Fee*	$5000 regardless of square footage			$500

Figure 2-5 *(Continues)*
LEED project certification rates.

	Less Than 50,000 sq. ft.	50,000–500,000 sq. ft.	More Than 500,000 sq. ft.	sq. ft. Appeals (If Applicable)
Construction Review				
USGBC Members	$500	$0.010/sf	$5000	$500
Non-Members	$750	$0.015/sf	$7500	$500
Expedited Fee*	$5,000 regardless of square footage			$500
Combined Design & Construction Review				
USGBC Members	$2250	$0.045/sf	$22,500	$500
Non-Members	$2750	$0.055/sf	$27,500	$500
Expedited Fee*	$10,000 regardless of square footage			$500
LEED for Existing Buildings	**Fixed Rate**	**Based on Square Footage**	**Fixed Rate**	**Per Credit**
Initial Certification Review				
USGBC Members	$1500	$0.03/sf	$15,000	$500
Non-Members	$2000	$0.04/sf	$20,000	$500
Expedited Fee*	$10,000 regardless of square footage			$500
Recertification Review**				
USGBC Members	$750	$0.015/sf	$7500	$500
Non-Members	$1000	$0.02/sf	$10,000	$500
Expedited Fee*	$10,000 regardless of square footage			$500
LEED for Core & Shell: Precertification	**Fixed Rate**			**Per Credit**
USGBC Members	$3250	$500		
Non-Members	$4250	$500		
Expedited Fee*	$5000	$500		
CIRs (for all Rating Systems)				$220

Figure 2-5 *(Continued)*

LEED project certification rates.

Please note that all fees are subject to change. No refunds are available.

* Project Square footage to be used for Certification Fee pricing should be based upon the definition of Gross Floor Area which is provided in the LEED 2009 MPR Supplemental Guidance. However, all parking areas (whether underground, structured, or at grade) should be excluded from the square footage calculations used to determine the certification fee. Â Other spaces such as common areas, mechanical spaces, and circulation should be included in the gross square footage of the building.

** In addition to regular review fee. Availability of expedited review timelines is limited based on GBCI capacity. Contact GBCI at least ten (10) business days prior to submitting an application to request an expedited review.

*** The Existing Building Recertification Review fee is due when the customer submits the application for recertification review. Before submitting, please contact GBCI's project certification staff to get a promotion code.

**** Precertification fees are for precertification only. Core & Shell projects are subject to all standard fees at the time of certification.

Step Two: Application Preparation

The second step in the certification process involves preparation of the certification application. This step is also undertaken online. The application is actually not a singular event but is a process that is undertaken over time as the project is designed and constructed. Each prerequisite or credit has a set of requirements that must be met in order to be awarded that particular prerequisite or credit. The project's compliance with these requirements must be proven through a unique set of documentation requirements. In some cases, a simple input of data is sufficient, for instance credit SS-3 Brownfield Redevelopment requires documentation that the site is a Brownfield site and a description of the remediation efforts used to clean up the site. However, in other cases detailed calculations must be submitted such as in the case of credit MR-4 Recycled Content, in which detailed calculations regarding the percentages of both pre-consumer and post-consumer recycled content of materials must be submitted. In every case, the required documentation is submitted through templates that are available online. These templates are completed by the various constituents of the project team. Responsibilities for completion range from those completed by the design team, that is, the architects and engineers, to those completed by the construction team. Samples of these templates are illustrated in the second section of this book.

Step Three: Application Submittal

The third step in the certification process involves the submittal of the application for review by the GBCI. There are two types of information that must be submitted for certification. One of these is information on the design of the building including site location, site, and building design information. The second type of information deals with compliance during the actual construction process. The project administrator can choose to submit the project design and construction documentation separately as the information is available or combined at the end of the project. By splitting the submission, the project administrator will be able to make design revisions to achieve credits prior to the construction. Once the building is constructed the anticipated design credits might be lost if not deemed to be in compliance with the LEED requirement.

Step Four: Application Review

The fourth step in the certification process involves a review by the GBCI of the completed application. During this review, points will either be awarded or denied by the GBCI. As stated earlier, design points can be awarded prior to the completion of the project if the project administrator has chosen to submit documentation during the design process. Construction-related points cannot be awarded until after the completion of construction and submission of all construction documentation.

If there is a question about specific requirements, the project administrator can, for a fee, request a credit interpretation. If the project team has been denied a point, it has three options. The project administrator can appeal the denied points, accept the denial, or the team can attempt additional points in other categories to make up for the lost points.

Step Five: Certification Award

The final step in the process is the actual award of certification. The GBCI will award a level of certification from "certified" to "platinum." This award will be commensurate with the number of points achieved by the project.

BENEFITS OF LEED BUILDING CERTIFICATION

There are numerous benefits to LEED certified buildings. These include both tangible benefits like reduced operating costs, and intangible benefits, such as providing a more healthy environment for workers. According to USGBC, the many benefits of LEED buildings can provide a savings in building energy consumption of between 24% and 50%. This is a substantial financial savings over a conventionally designed and constructed building. LEED certified buildings have experienced a 40% savings in water usage. Harmful greenhouse gasses such as CO_2 have been reduced by between 33% and 39% by employing sustainable building techniques. Finally LEED certified buildings have been known to reduce the solid waste entering the waste steam by 70%. In addition to these factors, LEED buildings have been shown to have other values over non-LEED buildings; such as higher occupancy rates, higher rents, and a higher return on the investment.

A detailed study of LEED certified buildings demonstrated that when compared with conventional buildings LEED buildings are:

- More energy efficient than conventional buildings. In fact LEED certified buildings are 25%–30% more energy efficient.
- LEED certified buildings generally have a lower peak energy consumption than conventional buildings (see Figure 2-6).

Figure 2-6

Reduced energy use in LEED buildings as compared with conventional buildings.

	Certified	Silver	Gold	Average (a)
Energy Efficiency above Code standard	18%	30%	37%	28%
On site Renewable Energy (b)	0%	0%	4%	0%
Green Power (c)	10%	0%	7%	6%
Total	28%	30%	48%	36%

© CENGAGE LEARNING 2012

(a) The term "average" refers to the average savings achieved from the three LEED Certification categories listed in the table.

(b) On site renewable energy is energy created on site from sources such as solar, wind, water, or geothermal sources.

(c) Green Energy is a renewable energy purchased for renewable energy sources located off site.

- The study found that LEED certified buildings are much more likely to generate on site energy than conventional buildings. The Energy and Atmosphere (EA-2) On Site Renewable Energy category has up to seven credit points available for providing "Green" energy on site. This would include the use of solar, wind, water, and geothermal energies to create power for the building.
- The study also found that a LEED certified building was more likely to purchase renewable energy than a conventional building. LEED buildings can achieve points for purchasing power from renewable energy sources. This requirement is defined in credit EA-6 Green Power. A 2-year renewable energy contract to provide at least 35% of the buildings electricity is required.

According to USGBC, there are currently 26,385 projects that have been registered for certification. In addition to these, over 4327 projects have already been certified. This accounts for over 385 million square feet of certified commercial building space currently constructed in the United States. The chart below illustrates the increasing interest in LEED certification since 2000 (see Figure 2-7).

Figure 2-7

Number of LEED projects from 2000 to 2009

Year	Certified Projects	Registered Projects
2000	3	8
2001	5	82
2002	21	151
2003	47	195
2004	117	354
2005	201	805
2006	319	1100
2007	543	4449
2008	982	8069
2009	2089	11,172
Total	4327	26,385

COURTESY OF U.S. GREEN BUILDING COUNCIL—LEED REFERENCE GUIDE

One example of a LEED Platinum building is the Proximity Hotel in Greensborough, North Carolina. It achieved 55 out of a possible 69 possible points in LEED v2.2 to gain a LEED Platinum certification, the highest level of LEED certification. This luxury hotel was designed and constructed to use 39% less energy and 34% less water than other comparable hotels. It has over 70 energy and environmental health enhancements, including the use of extensive day lighting in public spaces, 4000 sq. ft. of roof top solar panels that supply 60% of the hotel's hot water and geothermal powered refrigeration for the hotel's restaurants. Low VOC paints, stains, carpets, and interior materials were used throughout the hotel's interior (see Figure 2-8).

Figure 2-8

Proximity hotel.

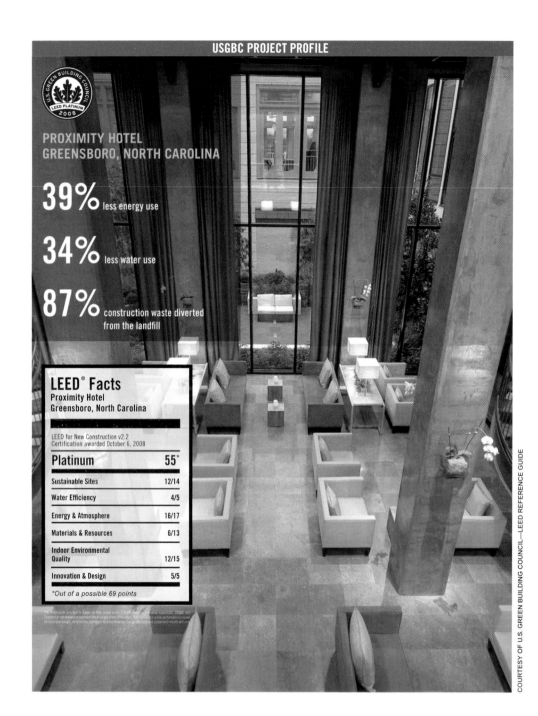

TAKE NOTE

Being a LEED constructor can lead to:
- More clients
- Increased profit
- Better buildings

LEED Benefits to the Contractor

Becoming a constructor of LEED certified buildings serves to demonstrate a contractor's commitment to environmental stewardship and social responsibility. Many contractors are using their "Green" approach to construction as a marketing tool to gain new clients. There are many clients who, if given the choice, would indicate that they are interested in constructing buildings that would have a reduced impact on the environment. Those clients will be looking for contractors with a similar philosophy. On the financial side, the constructors of LEED certified buildings are able to improve their "bottom line" through better C&D waste management. While this subject will be discussed in greater detail in Chapter 8, it is clear that some contractors have been able to substantially reduce the cost of C&D waste disposal on LEED projects. Lower costs always lead to increased profits.

THE COST OF LEED BUILDING CERTIFICATION

It has been commonly believed that the construction of a LEED certified building would be cost prohibitive in most cases. This idea could not be further from the truth. A study undertaken by Capital E involved the study of 30 "LEED" schools in 10 different states across the United States.[3] While the term "Green" does not always mean LEED certified, in this study 20 of the 30 schools were in fact LEED certified. The purpose of the study was to quantify the benefits is any of employing "Green" design and construction techniques in school construction. While only 20 of the 30 schools studied were in fact LEED certified the savings would be similar to LEED certified buildings. The results of the study showed that the cost of construction of "LEED Certified" schools as compared with conventional schools was much less than anticipated. In fact the additional initial construction cost was estimated at approximately 2% or $3.00 per square foot. But the measured benefits of building green far exceed the initial costs. As is illustrated by the following chart the study found that for that $3.00 per square foot initial investment a return of $71.00 per square foot was achievable (see Figure 2-9).

While it is true that not all of this money accrues directly to the school an approximate savings of $12.00 per square foot can be achieved in lower annual energy and water costs. The following chart illustrates the cost of making the schools "greener" as compared with the savings achieved. The standard used for these schools was the Massachusetts Collaborative for High Performance Schools (MA-CHPS).

Figure 2-9

Cost savings of building green.

	Cost Savings per Square Foot
Energy	$9.00
Emissions	$1.00
Water and Wastewater	$1.00
Increased Earnings	$49.00
Asthma Reduction	$3.00
Cold and Flu Reduction	$5.00
Teacher Retention	$4.00
Employment impact	$2.00
TOTAL Savings	$74.00
Cost of Greening	−$3.00
Net Financial Benefits	$71.00

© CENGAGE LEARNING 2012

The mission of the MA-CHPS is to facilitate the design, the construction, and the operation of high performance schools. These schools are intended to be not only energy and resource efficient, but also healthy, comfortable, and well-lighted (see Figure 2-10).

Name	State	Year Completed	2005 MA-CHPS	LEED Score	LEED Level or Equivalent	Cost Premium	Energy Savings	Wave Savings
Ash Creek Intermediate School	OR	2002			CERTIFIED	0.00%	30%	20%
Ashland High School*	MA	2005	19			1.19%	29%	
Berkshire Hills*	MA	2004	27			3.99%	34%	0%
Blackstone Valley Tech*	MA	2005	27			0.91%	32%	12%
Capuano	MA	2003		26	CERTIFIED	3.60%	41%	
Canby Middle School	OR	2006		40	GOLD	0.00%	47%	
Clackamas	OR	2002		33	SILVER	0.30%	38%	20%
Clearview Elementary	PA	2002	49	42	GOLD	1.30%	59%	39%
Crocker Farm School	MA	2001	37			1.07%	32%	62%
C-TBC	OH	2006	35	38	SILVER	0.53%	23%	45%
The Dalles Middle School	OR	2002			SILVER	0.50%	50%	20%
Danvers*	MA	2005	25			3.79%	23%	7%
Dedham*	MA	2006	32			2.89%	29%	78%
Lincoln Heights Elementary School	WA	2006			SILVER		30%	20%
Melrose Middle School	MA	2007	36			1.36%	20%	20%
Model Green School	IL	2004		34	SILVER	2.02%	29%	35%
Newton South High School	MA	2006		32	CERTIFIED	0.99%	30%	20%
Prairie Crossing Charter School	IL	2004		34	SILVER	3.00%	48%	16%
Punahou School	HI	2004		43	GOLD	6.27%	43%	50%
Third Creek Elementary	NC	2002		39	GOLD	1.52%	26%	63%
Twin Valley Elementary	PA	2004	41	35	SILVER	1.50%	49%	42%
Summerfield Elementary School	NJ	2006	42	44	GOLD	0.78%	32%	35%
Washington Middle School	WA	2006		40	GOLD	3.03%	25%	40%
Whitman-Hanson*	MA	2005	35			1.50%	35%	38%

Figure 2-10 *(Continues)*

Cost v. savings comparison of "Green" schools.

Williamstown Elementary School	MA	2002	37			0.00%	31%	
Willow School Phase 1	NJ	2003		39	GOLD		25%	34%
Wobum High School*	MA	2006	32			3.07%	30%	50%
Woodword Academy Classroom	GA	2002		34	SILVER	0.00%	31%	23%
Woodword Academy Dining	GA	2003		27	CERTIFIED	0.10%	23%	25%
Wrightsville Elementary School	PA	2003		38	SILVER	0.40%	30%	23%
AVERAGE						1.65%	33.4%	32.1%

Figure 2-10 (*Continued*)

Cost v. savings comparison of "Green" schools.

Another study undertaken by Capital E analysis of LEED certified office and school buildings resulted in similar results. In this 2003 study, LEED certified buildings had a premium cost for LEED certification of a negligible, 66%, whereas a LEED Platinum building incurred a cost premium of 6.5%. This premium is the percentage of construction cost per square foot above the average construction cost for a similar conventionally constructed building. The following chart illustrates the results of this study (see Figure 2-11).[4]

Figure 2-11

Average LEED cost premium *v.* level of LEED certification for offices and schools.

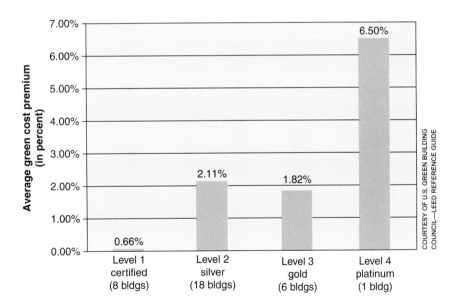

The same report concluded that the benefits of LEED buildings go far beyond that which was previously anticipated. As illustrated by the following chart the financial benefits of sustainable building design and construction are between $50 and $65 per square foot. This amount is a saving of over 10 times the premium associated with making the building LEED certified (see Figure 2-12).[5]

Financial Benefits of LEED Office Buildings	
Category	20 year Net Savings per SF
Energy Savings	$5.80
Emission Savings	$1.20
Water Savings	$0.50
Operations and Maintenance Savings	$8.50
Productivity and Health Benefits	$36.90–$55.30
Subtotal	$52.90–$71.30
Average Extra Cost of Green Construction	−$3.00–$5.00
Total 20-year Net Benefit	$50.00–$65.00

Figure 2-12

Financial benefits of LEED office buildings.

THE GREEN BUILDING CODE

The importance of "Green" construction and LEED building in particular is further illustrated by the announcement on March 11, 2010 of the nation's first set of model codes for green building. This new set of codes was developed jointly by The International Code Council (ICC), the American Society of Heating, Refrigeration and Air Conditioning Engineers (ASHRAE), USGBC, the Illuminating Engineering Society of North America (IES), the American Institute of Architects (AIA) and the American Society for Testing and Materials (ASTM). The new International Green Construction Code, IGCC, will contain much needed new content and language regarding sustainable construction. It will offer a new vision for a more safe and sustainable future.

THE CONTRACTOR'S ROLE IN LEED PROJECTS

THE LEED PROJECT TEAM

There are potentially seven primary participating entities in a construction project attempting to be LEED certified. They are as follows:

LEED Project Certification Team Members:

- The owner
- The LEED consultant. This person is hired by the owner to coordinate the LEED certification effort. They will become the LEED project administrators. If an independent consultant is not employed, the architect will generally become the LEED project administrator.
- The architect
- The landscape architect
- The design engineers, that is, Civil, Electrical, Plumbing, and Mechanical
- The general contractor
- The commissioning authority

The owners have the responsibility for taking the first step in deciding that the project they are planning will pursue LEED certification. In conjunction with a qualified LEED consultant, or the design architect, the owner will determine what level of certification will be attempted. The LEED consultant will be a LEED AP, generally having considerable experience in the LEED certification process. This consultant will be responsible for the determination of the owner and design team, for which credits will be attempted on the project to gain the specific LEED certification desired. The architects and their engineers will be responsible for designing the building, its components, and all electrical and mechanical systems to meet the requirements for LEED certification. An integral part of that design process will be the creation of a detailed set of specifications of "Green" materials and systems that must be complied with. Once the building has been designed and the construction documents produced, the responsibility passes on to the general contractor. While it is the general contractor's responsibility to assure that all of the requirements detailed in the construction documents are met by the construction team in order to make the required submittals, the contractor often receives information for other team members including the owner's representative (see Figure 3-1).

Figure 3-1

LEED project team relationships.

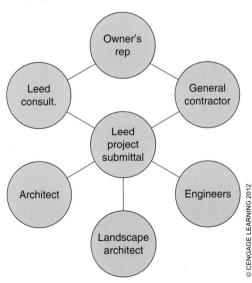

© CENGAGE LEARNING 2012

Commissioning Authority

The commissioning authority is an individual or firm who is responsible for assuring that the building systems are operating as designed. In LEED projects, it is the responsibility of a commissioning authority to ascertain that the building operating systems, air conditioning, plumbing and so on are meeting the LEED requirements for prerequisites or credits. There are two levels of commissioning involvement: fundamental commissioning, an Energy and Environment EA prerequisite and enhanced commissioning, a point earning category in EA-3 Enhanced Commissioning. The requirements for who can serve as a commissioning authority varies with the size of the building as well as with whether they will serve in the fundamental or enhanced commissioning operations. The following table illustrates who can perform as a commissioning authority of the LEED projects (see Figure 3-2).

Figure 3-2

Party acting as the commissioning authority.

Party Acting as the Commissioning Authority (CxA)	Fundamental Commissioning Prerequisite		Enhanced Commissioning Credit [3, 4; 5]
	<50k sq. ft.	>=50k sq. ft.	
1. Employee or subcontractor of the General Contractor with construction responsibilities	Y		
2. Employee or subcontractor, with construction responsibilities, of the Construction Manager (CM) who holds constructor contracts	Y		
3. Employee or subcontractor, with project design responsibilities, of the Architect or Engineer of Record (A/E)	Y		
4. Disinterested [1] employee or a sub-contractor of the General Contractor or CM	Y	Y	
5. Disinterested [1] employee of the A/E	Y	Y	
6. Disinterested [1] subcontractor to the A/E	Y	Y	Y
7. Construction Manager not holding constructor contracts	Y	Y	Y
8. Independent consultant contracted to Owner	Y	Y	Y
9. Owner employee or staff	Y	Y	Y

CONTRACTOR'S RESPONSIBILITY IN LEED PROJECTS

In the current version of the LEED requirements, as discussed in Chapter 2, there are seven credit categories. Each category contains a combination of prerequisites and credits that are to be fulfilled. The entity responsible for compliance with each prerequisite varies from credit to credit. The responsibilities fall on one of the following nine potential members of the LEED certification team; the owner, the architect, the landscape architect, the civil engineer, the mechanical engineer, the plumbing engineer, the electrical engineer, the contractor, and the commissioning authority.

The actual level of responsibility can vary from someone having primary responsibility for compliance to that of a supporting role in the LEED compliance process. For instance, the landscape architect would have the primary responsibility for compliance with credit WE 1.1, water-efficient landscaping reduce by 50%. The civil and plumbing engineers would have supporting roles in achieving compliance for that credit. While there is no specific mandate for who submits what information, the following chart illustrates a listing of the person typically responsible for submitting prerequisite and credit information for each LEED category (see Figure 3-3).

As can be seen from the chart above, while the contractors have no primary responsibility for prerequisite compliance, except if they act as a commissioning authority on projects under 50k sq. ft., they have secondary responsibility for the compliance of three of the six prerequisites, or 50% of the required prerequisites. In addition, they have the primary responsibility for the compliance of 9 credit categories, or 16% of all the possible credit categories and the secondary responsibility

Credit	Description	Owner	Architect	Landscape Architect	Civil Engineer	Mechanical Engineer	Plumbing Engineer	Electrical Engineer	Contractor	Commissioning Authority
Sustainable Sites										
Pre-1	Construction Activity Pollution Prevention			S	P				S	
Pre-2	Environmental Site Assessment								S	
SS-1	Site Selection	S		S	P					
SS-2	Development Density and Community Connectivity	P	S							
SS-3	Brownfield Development	S			P					
SS-4.1	Alternative Transportation Public Transportation Access		S		P					
SS-4.2	Alternative Transportation Bicycle Storage and Changing Rooms		S		P					

Figure 3-3 *(Continues)*

Responsibility for LEED credit compliance, where P and S denote primary responsibility and secondary responsibility, respectively.

SS-4.3	Alternative Transportation Low-Emitting & Fuel Efficient Vehicles		S		P						
SS-4.4	Alternative Transportation Parking Capacity		S		P						
SS-5.1	Site Development Protect or Restore Habitat		S	P	S						
SS-5.2	Site Development Maximize Open Space		S	P	S						
SS-6.1	Stormwater Design Quantity Control		S	S	P		S				
SS-6.2	Stormwater Design Quality Control		S	S	P		S				
SS-7.1	Heat Island Effect Non Roof		S	P	S						
SS-7.2	Heat Island Effect Roof		P								
SS-8	Light Pollution Reduction		S	S					P		
Water Efficiency											
Pre-1	Water Use Reduction	S			P	S					
WE-1	Water- Efficient Landscaping		S	P	S						
WE-2	Innovative Wastewater Technologies		S	S	S		P				
WE-3	Water Use Reduction		S	S			P				
Energy And Atmosphere											
Pre-1	Fundamental Commissioning of Building Energy Systems	S	S			S	S	S	S	S	P
Pre-2	Minimum Energy Performance		S			P		S			
EA-1	Optimize Energy Performance		S			P		S			S
EA-2	On-Site Renewable Energy		S			P		S			
EA-3	Enhanced Commissioning	S	S			S		S	S	P	
EA-4	Enhanced Refrigerant Management					P					
EA-6	Green Power	P									
Materials and Resources											
Pre-1	Storage and Collection of Recyclables	S	P								

Figure 3-3 *(Continues)*

Responsibility for LEED credit compliance, where P and S denote primary responsibility and secondary responsibility, respectively.

ID	Description	1	2	3	4	5	6	7	8
MR-1/1.1	Building Reuse- Maintain Existing Walls, Floors, and Roof		P/S					P/S	
MR-1.2	Building Reuse- Maintain Interior Nonstructural Elements		P/S					P/S	
MR-2	Construction Waste Management		S					S	
MR-3	Materials Reuse		P					P	
MR-4	Recycled Content		S					P	
MR-5	Regional Materials		S					P	
MR-6	Rapidly Renewable Materials		S					P	
MR-6/7	Certified Wood		S					P	
Indoor Environmental Quality									
Pre-1	Minimum Indoor Air Quality Performance				P				
Pre-2	Environmental Tobacco Smoke (ETS) Control	P			S				
IEQ-1	Outdoor Air Delivery Monitoring		S		P				S
IEQ-2	Increased Ventilation		S		P				
IEQ-3/3.1	Construction Indoor Air Quality Management Plan- During Construction				S			P	S
IEQ-3.2	Construction Indoor Air Quality Management Plan- Before Occupancy				S			P	S
IEQ-4.1	Low- Emitting Materials- Adhesives and Sealants		P		S	S		S	
IEQ-4.2	Low- Emitting Materials-Paints and Coatings		P					S	
IEQ-4.3	Low- Emitting Materials- Flooring Systems		P					S	
IEQ-4.4	Low- Emitting Materials- Composite Wood and Agrifiber Products		P					S	
IEQ-4.5	Low- Emitting Materials- Furniture and Furnishings		P					S	
IEQ-4.6	Low- Emitting Materials- Ceiling and Walls Systems		P					S	

Figure 3-3 *(Continues)*

Responsibility for LEED credit compliance, where P and S denote primary responsibility and secondary responsibility, respectively.

IEQ-5	Indoor Chemical and pollutant Source Control		P			S	S				
IEQ-6.1	Controllability of Systems-Lighting		P/S					P/S			
IEQ-6/6.2	Controllability of Systems-Thermal Comfort					P		S		S	
IEQ-7/7.1	Thermal Comfort-Design					P		S		S	
IEQ-7.2	Thermal Comfort-Verification	P				S				S	
IEQ-8.1	Daylight and Views-Daylight		P								
IEQ-8.2	Daylight and Views-Views		P								
Innovation In Design											
ID-1	Innovation In Design	The responsibility is dependent upon the scope of the innovation.									
ID-2	LEED® Accredited Professional	P									
IEQ-7/7.1	Thermal Comfort-Design					P		S		S	
IEQ-7.2	Thermal Comfort-Verification	P				S				S	
IEQ-8.1	Daylight and Views-Daylight		P								
IEQ-8.2	Daylight and Views-Views		P								
Innovation In Design											
ID-1	Innovation In Design	The responsibility is dependent upon the scope of the innovation.									
ID-2	LEED® Accredited Professional	P									

Figure 3-3 (*Continued*)

Responsibility for LEED credit compliance, where P and S denote primary responsibility and secondary responsibility, respectively.

for 11 other credits. In all, the contractor has some level of responsibility for the compliance of 36% of all possible credits. The following chart illustrates the areas of responsibility of the contractor in LEED compliance (see Figure 3-4).

CONTRACTOR'S DUTIES IN LEED PROJECTS

When engaged in a LEED project, the contractor must pay careful attention to all aspects of the LEED standard. The contractor's responsibilities involve five different operations. They include the following:

- Reading and understanding the specific LEED requirement
- Developing the action plan if required
- Fulfilling the requirement
- Preparing calculations to document compliance
- Uploading the documentation to the LEED online project Web site

Credit	Description	Report	Calculations	Contractor
Pre-1	Construction Activity Pollution Prevention	R		S
Pre-2	Environmental Site Assessment	R		S
Pre-1	Fundamental Commissioning Of Building Energy Systems	R		S
EA-3	Enhanced Commissioning	R		S
MR-1/1.1	Building Reuse- Maintain Existing Walls, Floors, and Roof		C	P/S
MR-1.2	Building Reuse- Maintain Interior Nonstructural Elements		C	P/S
MR-2	Construction Waste Management	R	C	S
MR-3	Materials Reuse		C	P
MR-4	Recycled Content		C	P
MR-5	Regional Materials		C	P
MR-6	Rapidly Renewable Materials		C	P
MR-6/7	Certified Wood		C	P
IEQ-3/3.1	Construction Indoor Air Quality Management Plan- During Construction	R		P
IEQ- 3.2	Construction Indoor Air Quality Management Plan- Before Occupancy	R		P
IEQ-4.1	Low- Emitting Materials- Adhesives and Sealants		C	S
IEQ-4.2	Low- Emitting Materials-Paints and Coatings		C	S
IEQ-4.3	Low- Emitting Materials- Flooring Systems	R		S
IEQ-4.4	Low- Emitting Materials- Composite Wood and Agrifiber Products	R		S
IEQ-4.5	Low- Emitting Materials- Furniture and Furnishings	R		S
IEQ-4.6	Low- Emitting Materials- Ceiling and Walls Systems	R		S

Figure 3-4

Contractor responsibility for LEED credit compliance.

Reading and Understanding the Requirement

The first operation involves the reading and understanding of all aspects of the LEED requirement. These prerequisite or credit discussions include the following:

- A chart listing the number of credits possible.
- A discussion of the intent of the credit.

- Benefits of the requirement and other related issues to consider.
- A listing of other related credits. The LEED program is a strong proponent of synergism between the various requirements.
- A summary of the referenced standards.
- A discussion of the timeline as it relates to when this requirement should be completed and the project team members involved.
- A discussion of the various calculations that must be performed including examples of similar sample calculations.
- Guidance as to how the project team must document compliance.
- A discussion of exemplary performance and regional variations.
- A listing of resources.
- The definitions of important terms related to the requirement.

Developing the Action Plan

It is often very beneficial for the contractor to develop an action plan. This action plan will include specific action steps that are required to complete the requirement. Similar to how a construction schedule allows a contractor to track his or her progress on a project, this action plan can be used to keep track of the progress made toward LEED compliance.

Fulfilling the Requirement

The third operation involves the fulfillment of the specific requirement. This often requires an action such as diverting waste from the landfill. The specific LEED requirement might also require certain planning measures. Several of the credits require the preparation of a plan such as the development of a waste management plan, required in LEED Credit MR-2.

Preparing Calculations

Many LEED requirements require the contractors to undertake calculations to document that they have indeed complied with the standard. One such calculation involves the calculation of how much construction waste is diverted from the landfill. The following is a sample of a waste diversion calculation (see Figure 3-5).

Submission of Compliance Documentation

The last operation would involve the contractor submitting the documentation online through the LEED online Web site. The process involves the contractor accessing the project saved on the LEED online Web site. The contractor then opens the specific template for the credit they are working on. The template clearly

Project: Johnson Limited Warehouse		Total Project Cost			$650,000	
		Total Material Cost			$250,000	
Product Name	Vendor	Product Cost	Post-Consumer Content %	Pre-Consumer Content %	Recycled Content Value (*)	Recycled Content Info Source
Structural Steel	Acme Steel	$45,000	10%	85%	$23,625	Steel Manuf.
Concrete Aggregate	Jones Rock	$18,000	35%		$6300	Concrete Manuf.
Particleboard	Universal Builders Supply	$5000		100%	$5000	Manuf.
Drywall	Donavan Drywall Supply	$21,000		75%	$15,750	Manuf.
Total of the Recycled Content value					$50,675	
Recycled Content as a Percentage of Total Construction Cost					20.27%	
Total Points Achieved					2	

© CENGAGE LEARNING 2012

Figure 3-5

Sample calculation form for recycled content. *Denotes use the procedure to calculate recycled content.

indicates the information to be input in an organized tabular format. Samples of LEED online forms are included in Chapters 5–16.

While contractors are generally very proficient in the area of construction operations, they are not always as strong as in the areas of documentation. A contractor undertaking a LEED project might wish to hire someone familiar with the LEED documentation requirements to oversee this component of LEED compliance operations.

GREEN IS LEAN

The EPA defines green building as: "Green building is the practice of creating structures and using processes that are environmentally responsible and resource-efficient throughout a building's life-cycle from location on the site to design, construction, operation, maintenance, renovation, and deconstruction. This practice expands and complements the classical building design concerns of economy, utility, durability, and comfort. Green building is also known as 'sustainable or high performance building.'"

In addition, green building is also lean building. Lean construction has been defined as "a way to design production systems to minimize waste of materials, time, and effort in order to generate the maximum possible amount of value."[1] Green construction embodies many of the philosophies of "lean construction."

By conserving materials, reusing materials whenever necessary, minimizing waste produced through the construction process, and recycling whenever possible, green construction is a very efficient construction system.

Green contractors, that is those contractors who on a daily basis, in all aspects of their work and business, consider the impact of their operations on the environment. They manage the entire construction process in a material- and time-efficient manner. This concept extends well beyond the jobsite. For instance, the green contractors will carefully manage their material procurement process. By scheduling the ordering and delivery of materials in larger packaged groups from a single supplier, the green contractor can possibly minimize packaging waste. For instance, the purchasing of miscellaneous building supplies for the same company that supplies larger building materials and have them all delivered together the green contractor can eliminate not only packaging materials but might be able to significantly eliminate a number of material deliveries. Each delivery has associated with it an inherent cost for labor, fuel, and so on. Not to mention the pollution-laden emissions from the delivery vehicles. By reducing the number of deliveries to the job site, the green contractor has effectively helped the environment. As stated in Chapter 2, in addition to help protect the environment there are tangible benefits to the contractor by working greener and leaner. These benefits include a potential for greater profits due to the benefits of reduced cost through more effective project management, and the ability to target market to individuals and companies.

For more information on lean construction go to http://www.leanconstruction.org.

THE GREEN OFFICE

It is the hope of this author that the green principles discussed in this book and elsewhere should begin to permeate the contractor's business. This includes both on site and off site operations. Much of this book is dedicated to discussing green construction techniques; however, to truly be effective the contractors must "green up" their entire operation. There are many possible ways contractors can "green up" their office. The following are a few suggestions.

Save Energy

The contractor should take steps to assure that the office operations are as energy efficient as possible. This can be accomplished in a variety of ways including the following:

- Use an economizer on the air conditioning system. In moderate climates the air conditioning system can be modified to use cooler outside air instead of cooling the air with refrigerant. This could result in a significant savings over the more moderate months.
- Use a programmable thermostat. Programmable thermostats when properly used can not only provide better temperature control but can also save energy; by making the air conditioner work more effectively.

- Revise your office lighting system. This involves reducing the general lighting in office spaces and increasing the amount of task lighting at specific work spaces. Task lighting is the provision of light fixtures at individual work stations. These task lights are controlled by the user of the workspace and can be turned off when not needed.
- Replace T12 fluorescent bulbs and magnetic ballasts with more energy-efficient T8 bulbs and electric ballasts.
- Turn off your computers at night. The use of screen savers is deceptive. Many offices leave their computers on all night. Computers are notorious energy hogs. Unless absolutely necessary for communications linking reasons all computers should be turned off at the end of the day.
- Install motion sensor switches in all interior spaces especially restrooms, lounges, and kitchens. These are special electrical switches that will turn off the lights if not inhabitants are sensed. Make sure these switches have manual overrides.
- Use Energy Star office equipment includes a variety of equipment ranging from computers to water coolers. Each piece of equipment that has earned the ENERGY STAR rating, helps save energy through special energy-efficient designs. These designs allow the specific equipment to operate efficiently at lower power levels than that of non-ENERGY STAR rated equipment. Another feature of ENERGY STAR equipment is that they will automatically reduce power levels when not in use. The following is a sample of the ENERGY STAR efficiency requirements for certain types of office equipment. A full listing can be viewed at http://www.energystar.gov (see Figures 3-6 and 3-7).

Figure 3-6

Product criteria for ENERGY STAR qualified computers.

Version 5.0 Energy Efficiency Requirements: *Effective July 1, 2009*

Product Type	Requirements
	• Category A: <= 148.0 kWh • Category B: <= 175.0 kWh • Category C: <= 209.0 kWh • Category D: <= 234.0 kWh
Desktops, Integrated Computers	*Note: Computers with more advanced memory, graphics, or hard drive capability may qualify for additional energy allowances (capability adjustments) above these amounts.* • Category A: <= 40.0 • Category B: <= 53.0 • Category C: <= 88.5
Notebooks and Tablets	*Note: Computers with more advanced memory, graphics, or hard drive capability may qualify for additional energy allowances (capability adjustments) above these amounts.* Off Mode: <= **2.0 W** Idle State:
Small-Scale Servers	• Category A: <= **50.0 W** • Category B: <= **65.0 W**

Figure 3-7

Product criteria for ENERGY STAR water coolers.

Equipment	Specification
Water Coolers	• Cold Only and Cook and Cold Bottled Units: < 0.16 kW-hours/day. • Hot and Cold Bottled Units: < 1.20 kW-hours/day.

Save Paper

Operating a paperless or near paperless office can result in a significant gain for the environment. It has been estimated that office workers use approximately 122,000 sheets of paper per year. The reduction in 1 ton of waste paper saves 3.3 cu. yd. in a land fill. The best practice is to print on both side of the sheet and always use recycled paper.

Reduce Travel

Employees should engage in travel outside their primary city of operation only when absolutely necessary. A significant amount of financial, time, and energy resources is consumed in unproductive travel. USA Today and American Express both estimate that the average cost of a domestic trip in the United States is over $1100.[2] And flight delays and long waits in security lines are becoming increasingly common. In addition, air travel has been shown to be a substantial source of the greenhouse emissions discussed earlier.

The use phone and video conferencing is becoming increasingly more popular. In many cases, it can be used as an efficient alternative in place of actual travel. In addition, employees are also more prone to illness after traveling. Studies have shown an increased incidence of illness during and after travel. This can result from an exacerbation of a pre-existing condition or the contraction of an airborne illness during the travel itself.[3] With phone and video conferencing your employees can stay at home and be considerably more productive.

Maintain a Healthy Indoor Office Environment

Replace furniture and finishes within the office with healthier more environmentally safe products.

- *Furniture.* Replace any furniture that has exposed particle board. This material gives off toxic gasses and is harmful to the interior office environment. There are a number of companies that market sustainable, environmentally safe office furniture.
- *Paint.* Paint the interior with low or no volatile organic compound (VOC) paint. Traditional paints that are free from gas toxic chemicals harmful to the offices' inhabitants. See Chapter 14 for low VOC paint products.
- *Flooring.* Replace the carpeting with either environmentally safe carpeting certified by the Carpet Institute or use an environmentally tile or wood product. See Chapter 15 for low VOC carpet products.

Office Supplies

Purchase environmentally friendly office supplies. Many office supply manufacturers offer a full line of environmentally friendly office supplies and equipment. These products range from recycled paper to biodegradable cleaning products and energy star equipment. Staples markets its environmentally friendly products under the name eco-easy, whereas Office Depot has similar products under the "Office deport Green Product" label.

Cleaning Products

It is understood that the cleaning products used in the cleaning of their office might be out of their control if these services are provided by their landlord. However, if the contractors do have control, they should purchase and use only environmentally safe cleaning products. Many manufacturers are now marketing residential and office cleaning products that are environmentally safe and contain no VOCs. These include the following:

- The Seventh Generation (http://www.seventhgeneration.com)
- Eco Concepts, Inc. (http://www.ecoconceptsusa.com)
- Sunshine Makers, Inc. (Simple Green) found at (http://www.simplegreen.com)

Recycle

Create a formal comprehensive recycling plan for the office. Recycle everything from paper products to aluminum cans. Provide conveniently located recyclable containers throughout the office. A good suggestion is to provide waste paper recycling containers at each workstation. The following items can be easily recycled:

- Paper
- Toner cartridges
- Printer ink cartridges

Green Up the Jobsite Trailer

The contractors should endeavor to maintain the same "Green" standards that they have for the main office in all of their remote job site trailers. This responsibility can be delegated to the project manager or highest level management authority on the jobsite. As the jobsite trailer is a relatively small space, the quality of the interior environment is of the utmost importance. Some of the steps to be taken include the following:

- Do not install finishes or furniture that emit VOCs. If the trailer comes pre-furnished try to limit the number of VOC emitting furniture wherever possible.

- Recycle all paper products.
- Recycle all toner and ink cartridges.
- Use only safe low VOC cleaning products. If the cleaning of the jobsite trailer is contracted out specify that the cleaning company use only "Green" products.
- Replace AC filters with a high MERV filter and inspect often. Chapters 15 and 16 of this book contain a more detailed discussion of AC filtration.

Put Someone in Charge

The contractor must designate someone within the company to develop, review, and enforce the company's "Green" policies. Bonuses can be established for the achievement of certain "Green" standards. An example would be the creation of a bonus for the office's "Green Czar" if certain levels of recycling are met. The following form can be used to keep track of the types and amounts of recycled materials in the office (see Figure 3-8).

Landmark Construction Annual Office Recycling Goals							
			GOAL	**ACTUAL**			
Recycled Item	Amount of materials	Units	Recycling Goal in %	Volume Recycled	% Recycled	Volume Discarded	% Discarded
Waste Paper	700	lbs	90%	650 lbs	92.8%	50 lbs	7.2%
Plastic Bottles	150	lbs	80%	125 lbs	83.3%	25 lbs	16.7%
Aluminum Cans	75	lbs	80%	65 lbs	86.6%	10 lbs	13.4%
Toner Cartridge	15	Each	100%	15	100%	0	0%
Ink Cartridge	12	Each	100%	12	100%	0	0%

© CENGAGE LEARNING 2012

Figure 3-8

Construction office recycling goals.

GREEN OFFICE POLICIES

If the contractors are serious about sustainability, they should prepare a "Green" policy manual for the office. In addition to outlining the company's general philosophy on sustainability, it should have tangible policy standards that can be monitored for compliance. The following is a sample of a "Green" office policy for a legal firm.

Weil, Gotshal & Manges LLP

Green Policy

Document Production & Management

The Going Green initiative seeks to (a) decrease our consumption of paper and (b) increase our usage of paper that contains recycled content.

Energy Reduction & Conservation

There are three programs on which our Going Green initiative will focus at the outset, including (a) ENERGY STAR-compliant office equipment; (b) ENERGY STAR-compliant facilities/offices; and (c) transportation and mass transit.

Hazardous Electronic Waste

Many types of electronic products used in the workplace contain hazardous substances like lead and mercury. When these products reach the end of their useful lives or become obsolete, some are considered hazardous waste. The firm will undertake various alternatives to disposing equipment in a responsible manner.

"Green Initiative" Policy Statement

Consistent with our firm's approach to social responsibility, Weil, Gotshal & Manges is "Going Green." The firm's green policy has two primary goals: (a) to lessen the law firm's impact on the environment and (b) to become a standard-bearer in the legal industry in promoting responsible stewardship toward the environment and its natural resources.

Weil Gotshal has built and maintained over the years a firm culture that prizes excellence not only in the practice of law, but also in discharging the firm's responsibility to the communities in which our lawyers and staff live and work. Our green policy is consistent with the firm's commitment to good corporate citizenship and best management practices.

The initiative will involve programs and policies in the following areas:

Document Production & Management

The Going Green initiative seeks to (a) decrease our consumption of paper and (b) increase our usage of paper that contains recycled content.

Energy Reduction & Conservation

There are three programs on which our Going Green initiative will focus at the outset, including (a) ENERGY STAR-compliant office equipment; (b) ENERGY STAR-compliant facilities/offices; and (c) transportation and mass transit.

Hazardous Electronic Waste

Many types of electronic products used in the workplace contain hazardous substances like lead and mercury. When these products reach the end of their useful lives or become obsolete, some are considered hazardous waste. The firm will undertake various alternatives to disposing equipment in a responsible manner.

Educational & Volunteer Opportunities

By partnering with organizations such as the EPA ENERGY STAR program and bar associations, among others, Weil Gotshal seeks to offer its employees educational and volunteer opportunities that promote energy and natural-resource conservation for both the home and office.

Carbon Offsets

In parallel with our efforts to decrease use of energy and other resources, Weil Gotshal has committed itself to move our firm toward carbon neutrality by purchasing carbon offsets. Specifically, Weil Gotshal will purchase offsets on the Chicago Climate Exchange in an amount equal to the amount of carbon emissions we produce through air travel by our lawyers and operation of our offices.

Vendor & Client Programs

Our Going Green initiative seeks to address the consumer choices we make across the breadth of our operations, from office supplies and equipment to commercial real estate. We also seek to participate in our clients' vendor programs, wherever possible.

Pro Bono Synergies

There is a tremendous opportunity for our firm to get involved more broadly, via pro bono, in environmentally conscious projects around the world.

Office "Green Committees"

A key component to our Going Green initiative will be the formation of "Green Committees" in every office. The Green Committee will be responsible for promoting, implementing, monitoring, and reporting the status and progress of each office's Going Green efforts.

SUSTAINABLE PROJECT MANAGEMENT

PROJECT DELIVERY METHODS

The phrase "project delivery method" refers to the various methods under which a contractor can operate with respect to acquiring and delivering a construction project. The two distinct phases inherent in this method are the project acquisition phase and the project delivery phase. The project acquisition phase deals with how a contractor acquires a project, that is, how is the contractor selected to undertake the project. This phase of the work ranges from the most common method, which is through a competitive bidding process, to a selection based on a negotiation between the owner and the contractor. There are four primary project delivery methods used in the construction industry. They are bid-build, negotiated, design-build, and construction management. The bidding documents prepared by the owner or his or her representative in the acquisition phase will generally specify which project delivery method will be required on the project. The contractor must then address all of the particular interdisciplinary relationship requirements involved in that method.

Bid-Build

The first method is the most traditional and has been successfully used for many years in the United States. In this method, the owner initiates the project and then hires the design professional, generally an architect or engineer. The design professional is responsible for developing the initial design concept. Once the design concept is approved, this entity is then responsible for preparing a set of detailed construction documents, including the drawings and specifications, fully describing the project. These construction documents are then distributed to multiple contractors who in turn conduct a detailed quantity survey and prepare a cost estimate of the project. The cost basis of this bid will generally be either a lump sum/fixed cost or some variation of a cost plus a fixed fee arrangement. Once completed, the estimate is submitted to the owner in the form of a bid for construction. The owner reviews the submitted bids and makes a final selection of which contractor will be chosen to undertake the project. Generally, the lowest bidder is selected unless that bid is considered unresponsive. An unresponsive bid is a bid that does not meet the specific requirements of the bid documents. For example, if the bid documents require that the bid include a breakdown of certain unit costs and the submitted bid does not contain this breakdown, the bid can be rejected by the owner. Once selected as the successful bidder, the contractor will be responsible for undertaking all construction operations required by the construction documents. This includes work undertaken by his or her own employees in addition to the work that is subcontracted out for others to perform. The following chart illustrates this process (see Figure 4-1).

A subcontractor is distinguished from a general or prime contractor by his or her contractual relationships. A constructor who is in contract with an owner is called a prime contractor. A contractor working for another contractor who, in turn, has a contract with an owner, that is, the prime contractor, is called a

Figure 4-1

Bid build sequence diagram.

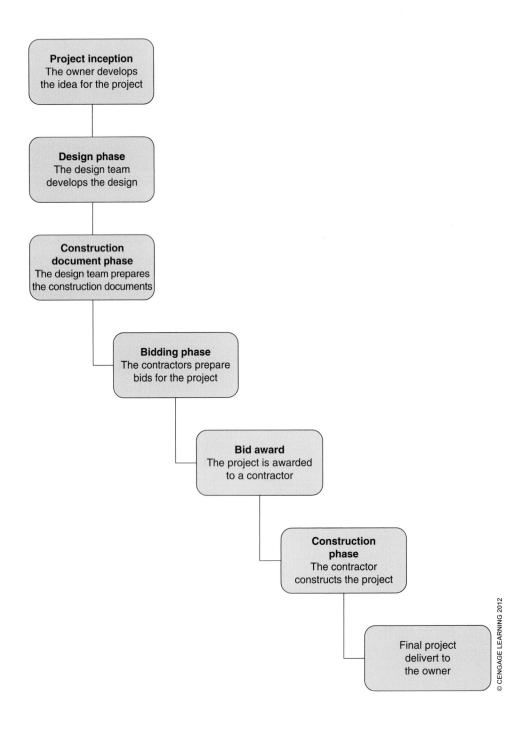

Project inception
The owner develops
the idea for the project

Design phase
The design team
develops the design

Construction
document phase
The design team prepares
the construction documents

Bidding phase
The contractors prepare
bids for the project

Bid award
The project is awarded
to a contractor

Construction
phase
The contractor
constructs the project

Final project
delivert to
the owner

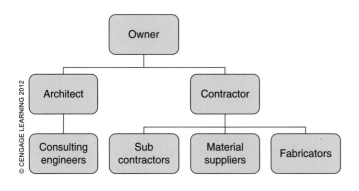

Figure 4-2

Traditional bid-build method of delivery.

subcontractor. The agreement which binds the two together is called a subcontract. This contract is similar in nature to the prime contract. In addition, subcontractors traditionally specialize in a narrow scope of work such as plumbing, electrical, mechanical, and so on. The contractor in this method of delivery will not be responsible for the actual, overall project design. The following diagram illustrates this methodology (see Figure 4-2).

Negotiated Contract Method

In this method, like the previously discussed bid-build method, the owner initiates the project and directly hires the design professionals. The design professional's responsibilities are the same as in the previous example. After the completion of the construction/bid documents, instead of submitting the project to several contractors for bids, only one contractor is approached. That one contractor is asked to prepare the estimate and bid to be submitted to the owner. This bid is then negotiated with the owner to achieve a final contract price. The basis of this cost, like in the previous project delivery method can be a fixed cost or a cost plus a fee. In the cost plus a fee, the contractor agrees to construct the project for a reimbursement of his or her costs plus a defined fee. This fee can be a fixed amount or can be a percentage of the reimbursable costs. Like the previous method, the contractor will have no responsibility for the design. The flow chart would be similar to the previous diagram illustrating the bid-build method.

Design-Build Method

In this method, as in the previous two methods, the owner initiates the project. But unlike the two previously discussed methods, instead of hiring the design professional directly, the owner contracts with the design-build contractor to provide both design and construction services. In this method, the owner and the contractor negotiate a budget within which the project will be designed and constructed. State laws regarding the practice of architecture and engineering would prohibit the contractor from undertaking the design by him or herself. Instead, the contractor, like the owner, would have to retain a design professional to design and produce the required construction documents. Occasionally, the design professional and the contractor are the same entity, and all design-build services are provided by a single company or corporation.

In concept, because the contractor has control of both the design and the construction processes, a number of issues often encountered in the two previous methodologies that result in expensive change orders during construction can be

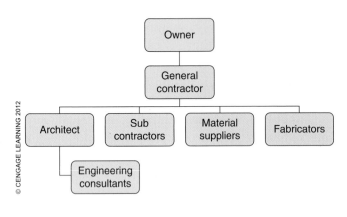

© CENGAGE LEARNING 2012

Figure 4-3

Design-build method of project delivery.

eliminated. In addition, because of the close working relationship of the design and construction teams, if allowable by the local building departments, the project can sometime be "fast tracked" with foundation construction being started prior to the total completion of the construction documents. Unlike the previous two examples, in this method of project delivery, the contractor is responsible for the design. The following is a diagram of the design-build project delivery method (see Figure 4-3).

Because of the potentially expedited project delivery, the design-build method has become a popular choice of many governmental agencies including local school boards, universities, and the U.S. Department of Transportation (USDOT). In fact according to a design-build effectiveness study undertaken by the USDOT, design-build method was shown to be a source of time and cost savings on a project while also potentially improving quality.[1] The study states that the design-build method can result in the following:

- **Time savings** through:
  - Early contractor involvement that enables construction engineering considerations to be incorporated into the design phase and enhances the constructability of the engineered project plans;
  - Fast-tracking of the design and construct portions of the project, with overlapping (concurrency) of design and construction phases for different segments of the project; and
  - Elimination of a separate construction contractor bid phase following completion of the design phase.
- **Cost savings** from:
  - Communication efficiencies and integration between design, construction engineering, and construction team members throughout project schedule;
  - Reduced construction engineering and inspection (CEI) costs to the contracting agency, that is, USDOT, when these quality control activities and risks are transferred to the design-builder;
  - Fewer change and extra work orders resulting from more complete field data and earlier identification and elimination of design errors or omissions that might otherwise show up during the construction phase;
  - Reduced potential for claims and litigation after project completion as issues are resolved by the members of the design-build team; and
  - Shortened project timeline that reduces the level of staff commitment by the design-build team and motorist inconvenience due to reduced lane closures.
- **Improved quality** through:
  - Greater focus on quality control and quality assurance through continuous involvement by design team throughout project development; and
  - Project innovations uniquely fashioned by project needs and contractor capabilities.

In addition to commercial construction, where this method of project delivery is flourishing, many residential contractors have associated themselves with local architects or residential designers to offer design-build services for homes. This gives them better control of project costs through a more comprehensive control on both the design and construction processes.

Construction Management

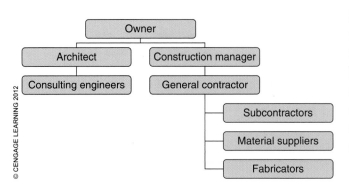

Figure 4-4

Construction management method of project delivery.

In this method, the owner retains a construction manager (CM) or construction management firm to act on their behalf. The CM responsibilities would be to act as the owner's representative and coordinate the entire design and construction processes (see Figure 4-4). This CM can either act "in agency," as an agent of the owner or can be "at risk."

CM Agency

If the CM is at agency, the CM would have no risk. In this method, the CM is often brought in during the design development phase. The CM can provide the design team with invaluable information and advice which can eventually result in reduced costs or reduced construction duration. Throughout the actual construction phase, the CM would be expected to use their best professional expertise and judgment in undertaking their duties. If there are problems encountered during construction, the CM would coordinate the finding of a solution but would not be held responsible for any losses. The exception would be if the losses were directly caused by the CM. In this event, the CM could be held responsible for the loss or damages.

CM at Risk

Sometimes, the owner will choose to place the CM at risk. If the CM is at risk, then the contract they have with the owner will have certain terms and conditions that will require the CM to deliver the project within a certain number of days and at a given cost. Construction managers at risk can incur damages because of noncompliance with the project contract requirements. The owner will generally determine which CM relationship works best for them. In some cases, the CM will start out as a CM agent but the relationship will evolve into a CM at risk during the construction phase. In any event, the function of the CM is to manage the entire construction process. The CM does not perform any of the work as they provide management services only.

TAKE NOTE

Four methods of project delivery:
- Bid-build
- Negotiated contract
- Design-construct
- Construction management

LEED PROJECT DELIVERY METHODS

The methods of delivery for LEED projects are generally similar to those discussed in the previous sections. Most of these methods of delivery could be effectively used on a LEED project. However, since the undertaking of a LEED project requires a substantial amount of close coordination between all members of the design and construction teams, the traditional bid-construct method would not be as effective as the design-build method. In the bid-build method, there is no coordination between the design and construction personnel until the construction begins. Often, the relation between the design and construction team is riddled with distrust and animosity. Allowing the contractor control of both the design and construction processes alleviates most of these issues. In the design-build method, a well-functioning, coordinated design construction team is in place at the inception of the project. This closely working team is certainly in the best position to deliver a project meeting all of the required LEED credits.

The primary difference in using the other delivery methods on a LEED project is the use of a LEED project administrator and a separate project commissioning authority. As discussed in Chapter 3, this LEED project administrator can be the design architect or can be an independent LEED consultant which is commonly brought into the project team. Unlike the architect who also has design and compliance responsibilities, this independent consultant will be a LEED accredited professional with generally considerable experience in the LEED certification process. The LEED consultant will have no specific design or individual compliance requirements but instead will be responsible for coordination of the LEED certification efforts of both the design and construction teams. The commissioning authority will be a person brought in to undertake the commissioning, that is, start-up of the building's electrical and mechanical systems. This entity has specific duties defined in the LEED standard.

Bid-Build on LEED Projects

On Bid-build projects, the owner in consultation with the LEED consultant will determine which level of certification, Certified, Silver, Gold, or Platinum, will be attempted and which credits can be achieved. Ultimately, the design architect and engineers will be responsible for the project design credits, whereas the commissioning authority, also discussed in Chapter 3, and the general contractor will have the prime responsibility for complying with all of the construction-related standards designated to be archived during the construction and commissioning of the project. As previously mentioned, this method, though often used, is not the most effective in undertaking a LEED compliant project.

The following is a typical organization of a LEED project team for a bid-build project (see Figure 4-5).

Figure 4-5

Bid-construct method of project delivery with LEED consultant.

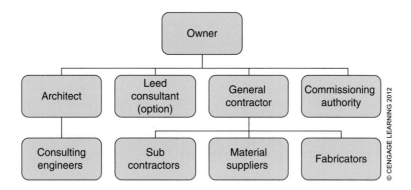

Design-Build on LEED Projects

The LEED team on a design build project can be similar to that of a traditional design build team with the addition of both the LEED consultant which is optional and the commissioning authority. In the case of design-build, the LEED Consultant and commissioning authority can be an independent consult to the owner as in Figure 4-6 or can be consultants to the general contractor as in Figure 4-7. The second option places more direct responsibility with the general contractor as general contractor is responsible for the entire LEED process. This is good for the owner who only looks to one source, the general contractor, for responsibility but not always so good for the general contractor because the general contractor will carry 100% of the liability for LEED compliance (see Figures 4-6 and 4-7).

Figure 4-6

Design-build method of project delivery with LEED consultant and commissioning authority hired by the owner.

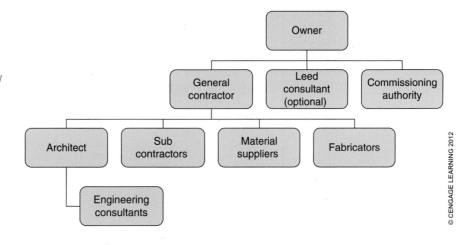

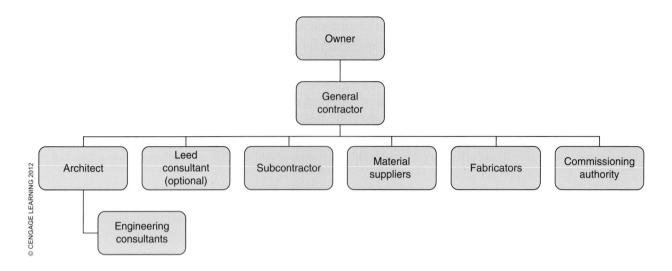

Figure 4-7

Design-build method of project delivery with LEED consultant and commissioning authority hired by the general contractor.

Construction Management on LEED Projects

The construction management method of project delivery leads itself well to LEED projects. After all, the CM is a "manager" so as such, is generally well-qualified to manage an interdisciplinary team. The addition of another two consultants can be independent consultants to the owner as in the bid-build organization, see Figure 4-8, or they can be consultants to the CM as in Figure 4-9. In this method of delivery, the CM will be involved as the overall project coordinator from the inception to the completion and project turnover to the owner. Having this overall coordination can lead to a successful completion of a LEED project (see Figures 4-8 and 4-9).

Figure 4-8

Construction management method of project delivery with LEED consultant and commissioning authority hired by the owner.

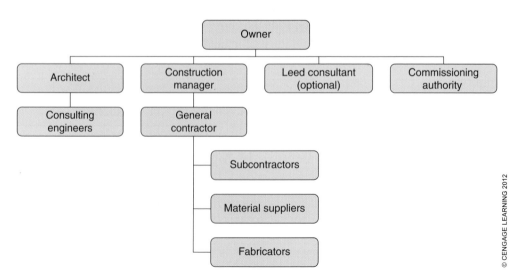

Figure 4-9

Design-build method of project delivery with LEED consultant and commissioning authority hired by the construction manager (CM).

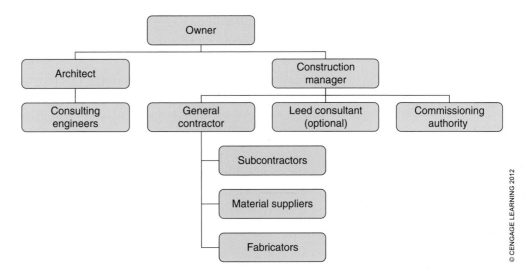

© CENGAGE LEARNING 2012

RISK IN LEED PROJECTS

There are very few industries in the United States that have the same level of risk as the construction industry. These risks exist because construction is a relatively long-term endeavor, often spanning several, if not more, years. It involves complicated processes with complex management situations. Some of the common risks faced in construction include the following:

- Insufficient or inadequate construction documents
- Uncertain site or existing conditions
- Weather
- Insufficient number and quality of available workers
- Labor or employee problems
- Delays in payments
- Insufficient credit or changes in available credit

In LEED projects, in addition to those risks identified above, there are additional risks. These additional risks are generally associated with LEED compliance issues. Unlike traditional building construction, projects attempting LEED certification require strict adherence to certain materials procurement guidelines and procedures.

Example

On a conventional construction project, the contractor can purchase a commonly found material such as drywall from any supplier. There would also be no restrictions on the composition of the drywall, except for thickness and fire rating. However, on a LEED project, the contractor will most likely have to purchase a type of drywall that contains a significant amount of recycled material. Then the contractor must carefully document that purchase, the percentage of recyclable materials, and compare them against the value of project materials. This documentation is described in more detail in Chapter 10.

Example

If an owner is developing an upscale office building, the developer's plan is to maximize rentals by marketing the building as a LEED "Platinum" building. Assume the building was properly designed to meet the requirements for LEED Platinum certification. If the contractor failed to properly comply with the requirements for one or more of the LEED points, the project could in fact fail to gain enough points to become LEED Platinum. If this was to occur, the owner might be able to quantify monetary damages that resulted from lowered rental rates caused by the reduction in certification from Platinum to Gold. If this was the case, the contractor who failed to perform up to the required LEED standard might incur considerable responsibility.

Since the loss of a single point can adversely affect certification, the contractor must not only meet the intent of the standard but also must meet every detail of the requirement. Failure of the contractor in his duty to comply with LEED requirements can cause the building to not be certified or be certified at a lower level than originally anticipated. This situation might lead to the owner attempting to sue the contractor for damages incurred because of a loss of certification.

Candice Rusie, an attorney with the firm of Gary D. Reeves Bodman, LLP, in an article written for the Michigan AGC, states that failure to comply with LEED requirements can create a substantial amount of liability for the contractor.

> For example, certain points under LEED mandate the use of green products or materials with the proper amount of recycled content, are locally produced, have obtained certain certifications, or otherwise meet defined green standards. If the specified green materials are not used in the specified ratios, GBCI will not award the project those points. This may result in a project obtaining certification at a lower level than expected, or failing to achieve certification altogether. An unattained certification can be devastating to a project owner, who may lose tax incentives, governmental funding, and potential purchasers or tenants who wanted a LEED certified building. Removing and replacing problematic material so that certification may be obtained may be costly at best and impossible at worst. This clearly opens the door for liability to fall on the contractor or subcontractor's head.[2]

The following article entitled Green Building Disputes Prompt Lawsuits, published in Environmentalleader.com illustrates a lawsuit involving LEED compliance on a building project.

> Lawsuits and claims are emerging as businesses embrace green building and other sustainable processes, says Harvey Berman, a partner at the law firm of Bodman LLP, in an article for the Ann Arbor Business Review.
>
> Berman said the first reported lawsuit involving a green building is believed to be a case filed in Maryland by a contractor against a developer. The developer alleged that the contract which included the project manual and specifications required that the contractor construct an environmentally sound green building in conformance with a Silver Certification Level according to U.S. Green Building Council's Leadership in Energy & Environmental Design (LEED) Rating System, said Berman.
>
> The contract documents attached to the developer's counterclaim contained no provisions limiting the contractor's liability relating to any green building aspects of the project, did not discuss the role of the developer or the design/construction team regarding the LEED certification and did not discuss the LEED process or the effect of a project delay on the attainment of tax credit but the contract did contain a construction time limit of 336 days.

Since most green building standards involve some type of incentive for building green, there is an inherent potential liability for design professionals and contractors if an owner does not achieve the required green certification, warned Berman.

Eventually, in 2008, the case was settled and while the terms of the settlement are not public; Berman says "it is safe to say that the contractor was not paid any money as part of the settlement because the potential liability arising out of this green building project far exceeded any potential gain that the contractor might otherwise be entitled to receive."[3]

Contractors must understand that if they operate in the design-build method on a LEED project, they will more than likely be totally responsible for LEED compliance. This will be the case in both failures in both the design and the construction aspects of LEED compliance.

TAKE NOTE

Five keys to managing risk:
1. Avoidance
2. Abatement
3. Transfer
4. Retention
5. Allocation

MANAGING RISK IN CONSTRUCTION PROJECTS

There are five major methods to manage risk on construction projects. They are as follows:

- Risk Avoidance
  A contractor can avoid any risk involved in a construction by simply deciding not to bid on the project.
- Risk Abatement
  This generally employs a combination of risk prevention and loss reduction techniques. The purpose of risk abatement is not to totally eliminate risk but to minimize the impact of risk on the contractor.
- Risk Transfer
  This is the transfer of risk to another party. This is generally accomplished through a traditional insurance policy.
- Risk Retention
  When the risk cannot be transferred away from the contractor or when the cost of that risk approaches the potential loss, the contractor will generally retain the risk.

Traditional project

An example of this might be the discovery that a contractor failed to install the proper type of "low VOC" insulation in the exterior walls. This would fall under Credit IEQ-4.6 Low-Emitting Materials—Ceiling and Walls Systems. If this error was discovered prior to the installation of the interior wall finish, the situation could be easily corrected. However, if this error was found after the wall finishes had been installed, a significant amount of time and money will have to be expended by the contractor to their construction error. This will often include the removal and replacement of both the insulation and the interior finishes. This process will often also result in significant delays that will also cost the contractor.

In projects attempting LEED certification, an error on the part of the contractor might preclude the award of certain anticipated credits. If these credits cannot be achieved in other ways, the project might not be able to achieve the desired level of certification. This particular type of error could create a significant legal issue for the contractor.

- Risk Allocation

    When the risk is substantial, the contractor can chose to share the risk with other parties. This can be accomplished by establishing a joint venture partnership with a certain percentage of the risk being allocated to each partner. The contractor can also share the risk with his or her subcontractors.

Each contractor must develop his or her own risk management program for every given project.

MANAGING RISK IN LEED PROJECTS

One of the keys to managing risk on a LEED construction project is the development of a thorough understanding of all of the interrelated requirements for the project. "What are these requirements?" and "where can they be found?" are two questions that must be answered by every contractor. Where are the requirements for any given project found? The project requirements are found in not just one, but in several different sources. These include the construction contract and general conditions, the construction drawings, the specifications, the applicable building codes, and sometimes they are found in other standards. On projects intended to be LEED certified, the specific LEED requirements can be found on the GBCI Web site at GBCI.org.

In order to manage the risk on any given project, it is imperative for the contractor to thoroughly read and understand all of the requirements and standards that the contractor will be expected to comply with. One of the greatest causes of problems commonly encountered on construction projects is the contractor's noncompliance with a given standard. On some occasions, the noncompliance is discovered early enough to make corrections without incurring a significant loss of time or money. However, if the area of noncompliance is discovered after large portions of the work have been completed, then the corrections can involve a significant loss of both time and money.

An error such as that discussed above might not always be the sole responsibility of the contractor. In the event that the designer specified a chemical fire resistance application to the wood prior to installation, the responsibility for this error would then fall on the design professional who mis-specified the material. The contractor does not generally have a duty to discover and correct errors made by the designer.

Generally speaking, contractors do not like to get involved in design decisions. They often believe that by staying out of the jurisdiction of the designer, they are better protected from liability. On Green projects, the contractor must give up this notion for the advancement and success of the project. A successful LEED project will more than likely result from a true team effort in which each member of the team places the success of the project first.

For instance, involving the representatives of manufacturers of certified products throughout the project, through an organized program of quality/LEED compliance assurance, will be extremely valuable in assuring the successful obtaining of the attempted LEED points.

THE CONTRACT AND GENERAL CONDITIONS

The primary requirements for the contractor's performance on a project are delineated in the owner–contractor agreement. The general conditions of the contract for construction succinctly define the contractor's responsibilities. Some of the general requirements that are contained in many construction contracts are especially relevant to the construction of LEED projects because they establish certain responsibilities of the contractor during the construction process. These responsibilities range from the duty to carefully review the construction documents to the duty to supervise the construction process. The following is a list of some of these requirements:

- *Review of Contract Documents and Field Conditions by Contractors.*
  This clause generally requires the contractor to "carefully study and compare the various drawings and other documents," including information provided by the owner, for the project. This statement requires the contractor to have a full understanding of the project through a thorough review of the interrelatedness of all of the construction documents prior to initiating construction operations.
  On LEED projects, the contractor must be able to identify all of the LEED related materials, procedures, and other requirements to assure LEED compliance. If during this review, the contractor identifies a problem with LEED compliance for some reason, they have a duty to bring the issue to the attention of the designer as soon as the problem is identified.
- *Supervision and Construction Procedures.*
  This clause often requires the contractor to "supervise and direct the work." It goes on to place the full responsibility for construction means and methods

Example

LEED project

Assume that the contractor was constructing a project that was targeting a certain LEED certification. To protect the quality of the indoor air environment, the design calls for no treated wood to be used anywhere in the project. For aesthetic reasons, the project has extensive exposed wood in the interior. Assume that during the latter stages of construction, the contractor realizes that all of the wood used in the interior had been treated by a chemical to increase its fire resistance and that this chemical is being released into the indoor environment. In addition to losing the points for that particular LEED credit, the indoor atmosphere has been contaminated, which might jeopardize other LEED points. Since the work has already been installed and the damage done, the corrective measures will be considerably costly.

on the contractor. This type of clause generally requires the contractor to have a "competent" supervisor on site at all times work is being performed.

On LEED projects, this supervision includes assuring that the LEED conforming materials and processes defined in the construction documents are in fact adhered during the construction. This supervision responsibility would also extend to the documentation requirements which are an integral part of LEED compliance. The "competent" supervision would have to be someone who is not only familiar with the construction process but also should have a basic understanding of the LEED requirements in order to effectively supervise a LEED project.

- *Warranty.*

This clause requires the contractor to warrant that the work performed will meet the requirements of the contract documents. If the work is found to not be in compliance, the contractor will be responsible for removing and replacing the noncompliant work. This clause can cause considerable risk for the contractor working on LEED projects. On these LEED projects, the requirement for LEED compliance would be an integral part of the contract documents. If the contractor fails to comply with these LEED requirements and that failure results in a loss of LEED certification or a reduction in the LEED certification level, the contractor might be liable for damages. In most cases, the warranty provided by the contractor under the contract would have a time limitation, generally 1 y, except for some equipment. While this time limit would apply to LEED items, most problems with noncompliance with LEED would be discovered immediately upon completion of the project, when the compliance documentation is reviewed by the GBCI.

- *Review and Approval of Submittals.*

This clause requires the contractor to review and approve all submittals and samples received from subcontractors or material suppliers, prior to submitting them to the architect. The contractor by approving these samples and submittals certify that they meet the requirements of the project documents. On LEED projects, the review and approval of submittals and samples is a significantly more complex operation than on traditional projects. On traditional projects, the contractor would be checking for size, color, and correctness of material—is it metal or plastic and so on. On LEED projects, although much of the material is specified by the designers, the contractor would still be responsible for checking a variety of material submittals from subcontractors and material suppliers for compliance with the specifications and the specific LEED required standards. This places the contractor in the position of having LEED competent people on staff to review the submittals and assure LEED compliance.

Generally speaking, on a LEED project, if a contractor approves a material or other submittal required to meet a certain requirement for LEED compliance and the material or submittal is later found to not be in compliance, the general contractor will carry a portion of the responsibility. This will be true even if others within the approval chain such as the Architect or Engineer approved the submittal. Each person would carry a portion of the liability.

- *Cleaning Up.*

  Most contracts have a cleanup clause. This clause requires the contractor to clean up the site and remove all waste and debris. The requirements for LEED certification take this assignment one step further. In traditional projects, how to dispose of construction waste and debris is generally not specified in any significant detail. There is generally a note requiring the contractor to comply with all applicable laws and regulations in the disposal of construction waste and debris. This is to prevent the contractor from dumping the debris illegally in an open field or other nonallowable area. In LEED projects, however, the disposal of construction waste and debris is carefully regulated. How the debris is to be collected, sorted, and recycled or disposed of is to be carefully documented. The contractor must complete a variety of LEED required forms and documents including precise calculations of weight and volume. See Chapter 8 for more information on LEED C&D waste management procedures.

- *Subcontractual Relations.*

  There is a clause in many contracts that requires the general contractor to require all subcontractors to be bound to all of the requirements of the contract documents as they apply to the subcontractor's specific discipline. This is often a difficult clause for the general contractor to enforce. On traditional projects, unless something goes wrong, the necessity to pass risk on does not matter. On LEED projects, there are many requirements for documentation that must be met by all participants on the construction team. This requires the subcontractors to be knowledgeable of the LEED standards and requirements for documentation for their respective disciplines. The general contractor must pass the responsibility for strict compliance to the subcontractors through the subcontract documents. This also means that the general contractor must not only consider cost and efficiency when selecting subcontractors but they must also consider a subcontractor's knowledge and experience with LEED compliance.

  One way a contractor can better assure compliance with LEED requirements from all subcontractors is to engage in one or more LEED training sessions prior to the commencement of construction. The purpose of these training sessions would be to make sure all contractors understand their responsibilities during the LEED compliance process.

- *Hazardous Materials.*

  Most contracts contain a clause that limits the contractor's ability to bring certain listed hazardous materials onto the job site. The standard form of

TAKE NOTE

Educate your subcontractors

An educated subcontractor is one of your most important assets on a LEED Project.

contract used on traditional construction projects lists certain chemicals and substances which are to be specifically banned from the construction site. If a contractor or subcontractor unwittingly brings the banned materials onto the site, remedial measures must be initiated as soon as practicable. On LEED projects, the presence of hazardous chemicals and substances on the project could permanently endanger the LEED certification of the project. Everyone on the project team must be aware of materials and substances that are not allowed on the construction site.

For example, one of the lawsuits previously discussed in this chapter centers on a subcontractor using a solvent-based cleaning agent during construction. The cleaner released a substantial quantity of VOCs into the building environment, which would not only endanger LEED compliance but in the example case actually made the occupants ill.

For example, if a subcontractor encountered an adhesive material that could not be readily removed from a finished wall surface, then the subcontractor might mistakenly bring a carbon tetrachloride cleaning agent into the building to remove the adhesive. Although very effective in its cleaning capability, this material will release a considerably high level of VOCs into the building's indoor environment. This might result in a much longer flush out period than originally anticipated. This could possibly result in an overall delay in turning over the building to the owner.

Another source of hazardous materials might be found in the existing building that is undergoing a major renovation. Hazardous materials such as lead paint and asbestos were commonly used in buildings of the past. Many older buildings contain such "banned" dangerous materials. Lead paint was banned from use in 1978 by the EPA. Paint used in houses and building constructed prior to that date have a high probability of containing led. Asbestos was commonly used in buildings from the 1930s to the mid-1970s. Buildings constructed during that time frame would have a high likelihood of containing asbestos. If these materials are encountered, it is imperative that the contractor stop work immediately and report to the design architect or engineer.

A third source of hazardous materials might be the site itself. On occasion, the history of use of a property might have gaps or long periods of time for which the site's specific use is not known. On a site such as this, it would be possible to encounter hazardous materials such as oil or other petroleum products which may have spilled into the ground over time. If contaminated soil is discovered during a foundation excavation, for instance, the contractor should stop the excavation and contact the project's design professional and the local office of the EPA.

- *Tests and Inspections.*

This often found clause places the responsibility for obtaining all tests and inspections, required by the contract documents or any governmental authority, squarely on the general contractor. These test and inspections commonly include building inspections, concrete testing, AC testing and balancing, and special structural inspections. The requirements on LEED projects are much more complicated. LEED projects require special procedures on air-conditioning equipment start up. It also requires additional testing and monitoring of air quality after building start up, also referred to as commissioning. These requirements must be fully understood and complied with by all on the construction team.

On LEED projects, a separate commissioning authority is required. As previously discussed, this entity can be hired directly by the owner or can be a consultant to the general contractor. This commissioning authority will have the responsibility for all building commissioning requirements of the LEED standard.

- *Termination by the Owner for Cause.*

Many contracts contain clauses that give the owner the right to terminate the contractor for a cause such as repeatedly failing to comply with the requirements of the project's contract documents. This is a standard clause on most construction contracts. It protects the owner by allowing him or her to terminate the contractor if the contractor repeatedly fails to comply with provisions within the contract documents. The owners on traditional construction projects are generally very reluctant to exercise this clause because of the myriad of legal consequences that may subsequently occur. However, on LEED projects, owners might exercise this option more quickly if they deem the contractor's repeated noncompliance is endangering the project's LEED certification. General contractors must be aware of the risks of failing to comply with the specific and unique requirements that are mandated when working on LEED projects.

Candice Rusie, an attorney with the firm of Gary D. Reeves Bodman, LLP says it best when she warns, in an article written for the Michigan AGC that. . .

Contractors should be wary if a contract promises the attainment of a certain green building certification, as under such a contract compliance directly depends upon a third party's determinations, made solely upon whatever documents were submitted to it. Contractors should also be wary of a contract relating payment to an ultimate result such as the successful award of certain LEED points or certification, rather than to the contractors' successful performance of specific services during construction. This is not only due to the unpredictable nature of LEED certification, but also because the LEED review and certification process may take a long time to complete and is dependent upon timely, complete submissions of documents—something out of the contractor's control. Similarly, contractors should provide warranties for their specific services only, and not guarantee the award of certain LEED points or LEED certification.[4]

TAKE NOTE

Contract requirements on LEED projects
- Review the construction documents
- Understand your responsibilities under the contract
- Educate the subcontractors

THE CONSTRUCTION DRAWINGS

The second place that the contractor would find requirements for the project would be in the construction drawings. The construction drawings traditionally contain all of the quantitative information for the project. However, notes contained in the drawings can sometimes offer a substantial amount of qualitative information. Design professionals commonly place a significant amount of construction quality information within notes placed in the construction drawings. On LEED projects, the contractor must care review these notes for specific LEED compliance information. For instance, the following Figure 4-10 is a typical concrete masonry note that is found on a structural sheet of conventionally designed project. Figure 4-11 illustrates a similar concrete masonry note that might be found on a LEED project. The LEED project CMU note contains much more detail regarding compliance with LEED requirements (see Figures 4-10 and 4-11).

Figure 4-10

Concrete masonry note on a traditional project.

MASONRY WALLS AND PARTITIONS

CONCRETE MASONRY UNITS (BLOCK) SHALL COMPLY WITH THE PROVISIONS OF THE STANDARD SPECIFICATION FOR THE DESIGN AND CONSTRUCTION OF LOAD BEARING CONCRETE MASONRY, NCMA TR–75B OR ACI 531.

HOLLOW BLOCK SHALL COMPLY TO ASTM C–90. TYPE I, GRADE N–1. DOUBLE END CELL BLOCK REQUIRED.

MORTAR SHALL COMPLY WITH ASTM C–270, TYPE M, WITH A MINIMUM COMPRESSIVE STRENGTH AT 28 DAYS OF 2500 PSI.

GROUT SHALL CONFORM TO ASTM C476, FOR COARSE GROUT, MINIMUM COMPRESSIVE STRENGTH © 28 DAYS OF 3000 PSI. SLUMP OF 8 TO 10 INCHES, MAXIMUM PLACEMENT TIME OF 90 MINUTES. SUBMIT DESIGN MIX FOR APPROVAL.

HORIZONTAL REINFORCING SHALL BE DUR–O–WALL STANDARD (9 GA.) TRUSS, ASTM CLASS B–2, HOT DIPPED GALVANIZED OR APPROVED EQUAL. PLACE VERTICALLY AT ALTERNATE COURSES. USE PREFABRICATED CORNERS AND TEES AT WALL INTERSECTION. OVERLAP DISCONTINUOUS ENDS A MINIMUM OF 12″.

VERTICAL REINFORCING SHALL CONFORM TO ASTM A615, GRADE 60. FILL ALL REINFORCED CELLS WITH 3000 PSI GROUT AS SPECIFIED ABOVE. SEE PLAN FOR SIZE AND SPACING OF VERTICAL REINFORCING.

MASONRY COMPRESSIVE STRENGTH f'm = 1500 PSI (COMPRESSIVE STRENGTH OF MASONRY = 1900 PSI)

MASONRY WALLS AND PARTITIONS

CONCRETE MASONRY UNITS (BLOCK) SHALL COMPLY WITH THE PROVISIONS OF THE STANDARD SPECIFICATION FOR THE DESIGN AND CONSTRUCTION OF LOAD BEARING CONCRETE MASONRY, NCMA TR–758 OR ACI 531.

HOLLOW BLOCK SHALL COMPLY TO ASTM C–90, TYPE I, GRADE N–1. DOUBLE END CELL BLOCK REQUIRED.

ALL MASONRY UNITS SHALL CONTAIN A MINIMUM OF 20% OF RECYCLED MATERIALS IN COMPLIANCE WITH LEED MR CREDIT 4.

THE MATERIALS FOR ALL CONCRETE MASONRY UNITS MUST BE OBTAINED FROM SOURCES WITHIN A 500 MILE RADIUS FROM THE SITE IN COMPLIANCE WITH LEED MR CREDIT 5.

MORTAR SHALL COMPLY WITH ASTM C–270. TYPE M. WITH A MINIMUM COMPRESSIVE STRENGTH AT 28 DAYS OF 2500 PSI.

MORTAR FOR MASONRY SHALL CONTAIN A MINIMUM OF 20% OF RECYCLED MATERIALS IN COMPLIANCE WITH LEED MR CREDIT 4.

THE MATERIALS FOR THE CONCRETE MASONRY UNIT MORTAR MUST BE OBTAINED FROM SOURCES WITHIN A 500 MILE RADIUS FROM THE SITE IN COMPLIANCE WITH LEED MR CREDIT 5.

GROUT SHALL CONFROM TO ASTM C475, FOR COARSE GROUT, MINIMUM COMPRESSIVE STRENGTH © 28 DAYS OF 3000 PSI. SLUMP OF 8 TO 10 INCHES, MAXIMUM PLACEMENT TIME OF 90 MINUTES. SUBMIT DESIGN MIX FOR APPROVAL.

GROUT FOR MASONRY SHALL CONTAIN A MINIMUM OF 20% OF RECYCLED MATERIALS IN COMPLIANCE WITH LEED MR CREDIT 4.

THE MATERIALS FOR THE CONCRETE MASONRY UNIT GROUT MUST BE OBTAINED FROM SOURCES WITHIN A 500 MILE RADIUS FROM THE SITE IN COMPLIANCE WITH LEED MR CREDIT 5.

HORIZONTAL REINFORCING SHALL BE DUR–O–WALL STANDARD (9 GA.) TRUSS, ASTM CLASS B–2, HOT DIPPED GALVANIZED OR APPROVED EQUAL PLACE VERTICALLY AT ALTERNATE COURSES. USE PREFABRICATED CORNERS AND TEES AT WALL INTERSECTION. OVERLAP DISCONTINUOUS ENDS A MINIMUM OF 12″.

STEEL FOR MASONRY SHALL CONTAIN A MINIMUM OF 20% OF RECYCLED MATERIALS IN COMPLIANCE WITH LEED MR CREDIT 4.

VERTICAL REINFORCING SHALL CONFORM TO ASTM A615, GRADE 60. FILL ALL REINFORCED CELLS WITH 3000 PSI GROUT AS SPECIFIED ABOVE. SEE PLAN FOR SIZE AND SPACING OF VERTICAL REINFORCING.

MASONRY COMPRESSIVE STRENGTH f'm = 1500 PSI (COMPRESSIVE STRENGTH OF MASONRY = 1900 PSI)

Figure 4-11

Concrete masonry note on a LEED project.

THE SPECIFICATIONS

The third place the contractor will find information on the requirements for a project is in the project specifications. The specifications is a written document which describes all of the qualitative aspects of the project. It defines the qualities of both the materials and the level of workmanship required for the manufacturing and installations of these materials.

Much of the LEED compliance information will be contained in the specifications. Rather than distributing the LEED requirements throughout the specifications as applicable to the various sections, many specifications for LEED projects contain a comprehensive LEED section. This section is generally found in CSI Division 1 General Requirements Section 01352. It would specify in detail what LEED compliance methodologies were being undertaken in that particular project. The following is an example of a comprehensive LEED Requirement specification.

CONSTRUCTION MASTER SPECIFICATION

SECTION 01350

LEED REQUIREMENTS

CONSTRUCTION MASTER SPECIFICATION

SECTION 01350

<u>LEED REQUIREMENTS</u>

PART 1 - <u>GENERAL</u>

1.01 Summary

    A. Section Includes: This Section includes general requirements and procedures for compliance with certain U.S. Green Building Council's (USGBC) <u>Leadership in Energy and Environmental Design</u> (LEED) prerequisites and credits needed for the Project to obtain LEED **[Certified]** **[Silver]** **[Gold]** **[Platinum]** certification.

        1. Other LEED prerequisites and credits needed to obtain LEED certification are dependent on material selections and may not be specifically identified as LEED requirements. Compliance with requirements needed to obtain LEED prerequisites and credits may be used as one criterion to evaluate substitution requests.

        2. Additional LEED prerequisites and credits needed to obtain the indicated LEED certification are dependent on the Architect's design and other aspects of the Project that are not part of the Work of the Contract.

    B. Related Sections: Refer to the following sections for related work:

        1. Divisions 1 through 16 Sections for LEED requirements specific to the Work of each of those Sections. These requirements may or may not include reference to LEED.

1.02 REFERENCES

    A. Code of Federal Regulations

        1. 40 CFR 59, Subpart D

    B. Forest Stewardship Council

        1. FSC 1.2, "Principles and Criteria"

    C. American Society of Heating, Refrigeration, and Air Conditioning Engineers

        1. ASHRAE 52.2.

    D. Sheet Metal and Air Conditioning Contractors' National Association

        1. IAQ Guideline for Occupied Buildings under Construction.

    E. Environmental Protection Agency

        1. Protocol for Environmental Requirements, Baseline IAQ and Materials, for Research Triangle Park Campus, Section 01445.

1.03 DEFINITIONS

    A. Certificates of Chain-of-Custody: Certificates signed by manufacturers certifying that wood used to make products was obtained from forests certified by an FSC-accredited certification body to comply with FSC 1.2, "Principles and Criteria." Certificates shall include evidence that mill is certified for chain-of-custody by an FSC-accredited certification body.

    B. LEED: Leadership in Energy & Environmental Design

C. Rapidly Renewable Materials: Materials made from agricultural products that are typically harvested within a ten-year or shorter cycle. Rapidly renewable materials include products made from bamboo, cotton, flax, jute, straw, sunflower seed hulls, vegetable oils, or wool.

D. Regionally Manufactured Materials: Materials that are manufactured within a radius of 500 miles (800 km) from the Project location. Manufacturing refers to the final assembly of components into the building product that is installed at the Project site.

E. Regionally Extracted, Harvested, or Recovered Materials: Materials that are extracted, harvested, or recovered and manufactured within a radius of 500 miles (800 km) from the Project site.

F. Recycled Content: The percentage by weight of constituents that have been recovered or otherwise diverted from the solid waste stream, either during the manufacturing process (pre-consumer), or after consumer use (post-consumer).

    1. Spills and scraps from the original manufacturing process that are combined with other constituents after a minimal amount of reprocessing for use in further production of the same product are not recycled materials.

    2. Discarded materials from one manufacturing process that are used as constituents in another manufacturing process are pre-consumer recycled materials.

1.04 SUBMITTALS

A. General: Submit additional LEED submittal requirements included in other sections of the Specifications.

B. LEED submittals are in addition to other submittals. If submitted item is identical to that submitted to comply with other requirements, submit duplicate copies as a separate submittal to verify compliance with indicated LEED requirements.

C. Project Materials Cost Data: Provide statement indicating total cost for building materials used for Project. Include statement indicating total cost of mechanical and electrical components.

D. LEED Action Plans: Provide preliminary submittals within 14 days of date established for commencement of the work indicating how the following requirements will be met.

    1. Credit MR 2.1 and 2.2: Waste management plan complying with Division 1 Section "Construction Waste Management."

    2. Credit MR 3.1 and 3.2: List of proposed salvaged and refurbished materials.

        a. Identify each material that will be salvaged or refurbished materials.

    3. Credit MR 4.1 and 4.2: List of proposed materials with recycled content.

        a. Indicate cost, post-consumer recycled content, and pre-consumer recycled content for each product having recycled content.

    4. Credit MR 5.1 and 5.2: List of proposed regionally manufactured materials and regionally extracted, harvested, or recovered materials.

        a. Identify each regionally manufactured material, its source, and cost.

        b. Identify each regionally extracted, harvested or recovered material, its source, and cost.

     5. Credit MR 7.0: List of proposed certified wood products.

       a. Indicate each product containing certified wood, its source, and cost.

       b. Include statement indicating total cost for wood-based materials used for Project, including non-rented temporary construction.

     6. Credit EQ 3.1: Construction indoor air quality management plan.

E. LEED Progress Reports: Concurrent with each Application for Payment, submit reports comparing actual construction and purchasing activities with LEED action plans for the following:

     1. Credit MR 2.1 and 2.2: Waste reduction progress reports complying with Division 1 Section "Construction Waste Management."

     2. Credit MR 3.1 and 3.2: Salvaged and refurbished materials.

     3. Credit MR 4.1 and 4.2: Recycled content.

     4. Credit MR 5.1 and 5.2: Regionally manufactured materials and regionally extracted, harvested, or recovered materials].

F. LEED Documentation Submittals:

     1. Credit SS 7.2: Product Data for roofing materials indicating Energy Star compliance.

     2. Credit SS 8.0: Product Data for interior and exterior lighting fixtures that stop direct-beam illumination from leaving the building site.

     3. Credit WE 2.0 3.1 and 3.2: Product Data for plumbing fixtures indicating water consumption.

     4. Prerequisite EA 3.0: Product Data for new HVAC equipment indicating absence of CFC refrigerants. Phase-out plan to replace CFC refrigerants in HVAC&R systems with CFC-free refrigerants within the Construction Period.

     5. Credit EA 4.0: Product Data for new HVAC equipment indicating absence of HCFC refrigerants, and for clean-agent fire-extinguishing systems indicating absence of HCFC and Halon.

     6. Credit EA 5.0: Product Data and wiring diagrams for sensors and data collection system used to provide continuous metering of building energy and water consumption performance over time.

     7. Credit MR 2.1 and 2.2: Comply with Division 1 Section "Construction Waste Management.

     8. Credit MR 3.1 and 3.2: Receipts for salvaged and refurbished materials used for Project, indicating sources and costs for salvaged and refurbished materials.

     9. Credit MR 4.1 and 4.2: Product Data and certification letter indicating percentages by weight of post-consumer and pre-consumer recycled content for products having recycled content. Include statement indicating costs for each product having recycled content.

    10. Credit MR 5.1 and 5.2: Product Data indicating location of material manufacturer for regionally manufactured materials.

       a. Include statement indicating cost and distance from manufacturer to Project for each regionally manufactured material.

       b. Include statement indicating cost and distance from point of extraction, harvest, or recovery to Project for each raw material used in regionally manufactured materials.

11. Credit MR 6.0: Product Data for rapidly renewable materials.
    a. Include statement indicating costs for each rapidly renewable material.
12. Credit MR 7.0: Product Data and certificates of chain-of-custody for products containing certified wood.
    a. Include statement indicating costs for each product containing certified wood.
    b. Include statement indicating total cost for wood-based materials used for Project, including non-rented temporary construction.
13. Credit EQ 1.0: Product Data and Shop Drawings for carbon dioxide monitoring system.
14. Credit EQ 3.1:
    a. Construction indoor air quality management plan.
    b. Product Data for temporary filtration media.
    c. Product Data for filtration media used during occupancy.
    d. Construction Documentation: Six photographs at three different occasions during construction along with a brief description of the SMACNA approach employed, documenting implementation of the IAQ management measures, such as protection of ducts and on-site stored or installed absorptive materials.
15. Credit EQ 3.2:
    a. Signed statement describing the building air flush-out procedures including the dates when flush-out was begun and completed and statement that filtration media was replaced after flush-out.
    b. Product Data for filtration media used during flush-out and during occupancy.
    c. Report from testing and inspecting agency indicating results of IAQ testing and documentation showing conformance with IAQ testing procedures and requirements.
16. Credit EQ 4.1: Product Data for adhesives and sealants used on the interior of the building indicating VOC content of each product used. Indicate VOC content in g/L calculated according to 40 CFR 59, Subpart D (EPA method 24).
17. Credit EQ 4.2: Product Data for paints and coatings used on the interior of the building indicating chemical composition and VOC content of each product used. Indicate VOC content in g/L calculated according to 40 CFR 59, Subpart D (EPA method 24).
18. Credit EQ 4.3: Product Data for carpet products indicating VOC content of each product used.
19. Credit EQ 4.4: Product Data for composite wood and agrifiber products indicating that products contain no urea-formaldehyde resin.
    a. Include statement indicating adhesives and binders used for each product.
20. Credit EQ 6.2: Product Data and Shop Drawings for sensors and control system used to provide individual airflow and temperature controls for minimum 50 percent of non-perimeter, regularly occupied space.
21. Credit EQ 7: Product Data and Shop Drawings for sensors and control system used to monitor and control room temperature and humidity.

PART 2 - PRODUCTS

    2.01   SALVAGED AND REFURBISHED MATERIALS

        A. Credit MR 3.1 and 3.2: Provide salvaged or refurbished materials for 10 percent of building materials (by cost). The following materials may be salvaged or refurbished materials:

        B. Credit MR 3.1 and 3.2: The following materials shall be salvaged or refurbished materials:

           1. Insert list of materials.

    2.02   RECYCLED CONTENT OF MATERIALS

        A. Credit MR 4.1: Provide building materials with recycled content such that post-consumer recycled content constitutes a minimum of five percent of the cost of materials used for the Project or such that post-consumer recycled content plus one-half of pre-consumer recycled content constitutes a minimum of 10 percent of the cost of materials used for the Project.

    2.03   REGIONAL MATERIALS

        A. Credit MR 5.1: Provide 20 percent of building materials (by cost) that are regionally manufactured materials.

        B. Credit MR 5.2: Of the regionally manufactured materials required by Paragraph "Credit MR 5.1" above, provide at least 50 percent (by cost) that are regionally extracted, harvested, or recovered materials.

    2.04   CERTIFIED WOOD

        A. Credit MR 7.0: Provide a minimum of 50 percent (by cost) of wood-based materials that are produced from wood obtained from forests certified by an FSC-accredited certification body to comply with FSC 1.2, "Principles and Criteria."

           1. Wood-based materials include but are not limited to the following materials when made from made wood, engineered wood products, or wood-based panel products:

              a. Rough carpentry.
              b. Miscellaneous carpentry.
              c. Heavy timber construction.
              d. Wood decking.
              e. Metal-plate-connected wood trusses.
              f. Structural glued-laminated timber.
              g. Finish carpentry.
              h. Architectural woodwork.
              i. Wood paneling.
              j. Wood veneer wall covering.
              k. Wood flooring.
              l. Wood lockers.
              m. Wood cabinets.
              n. Non-rented temporary construction, including bracing, concrete formwork, pedestrian barriers, and temporary protection.

2.05 LOW EMITTING MATERIALS

A. Credit EQ 4.1: For interior applications use adhesives and sealants that comply with the following limits for VOC content when calculated according to 40 CFR 59, Subpart D (EPA method 24):

1. Wood Glues: 30 g/L.
2. Metal to Metal Adhesives: 30 g/L.
3. Adhesives for Porous Materials (Except Wood): 50 g/L.
4. Subfloor Adhesives: 50 g/L.
5. Plastic Foam Adhesives: 50 g/L.
6. Carpet Adhesives: 50 g/L.
7. Carpet Pad Adhesives: 50 g/L.
8. VCT and Asphalt Tile Adhesives: 50 g/L.
9. Cove Base Adhesives: 50 g/L.
10. Gypsum Board and Panel Adhesives: 50 g/L.
11. Rubber Floor Adhesives: 60 g/L.
12. Ceramic Tile Adhesives: 65 g/L.
13. Multipurpose Construction Adhesives: 70 g/L.
14. Fiberglass Adhesives: 80 g/L.
15. Structural Glazing Adhesives: 100 g/L.
16. Wood Flooring Adhesive: 100 g/L.
17. Contact Adhesive: 250 g/L.
18. Plastic Cement Welding Compounds: 350 g/L.
19. ABS Welding Compounds: 400 g/L.
20. CPVC Welding Compounds: 490 g/L.
21. PVC Welding Compounds: 510 g/L.
22. Adhesive Primer for Plastic: 650 g/L.
23. Sealants: 250 g/L.
24. Sealant Primers for Nonporous Substrates: 250 g/L.
25. Sealant Primers for Porous Substrates: 775 g/L.

B. Credit EQ 4.2: For interior applications use paints and coatings that comply with the following limits for VOC content when calculated according to 40 CFR 59, Subpart D (EPA method 24) and the following chemical restrictions:

1. Flat Paints and Coatings: VOC not more than 50 g/L.
2. Non-Flat Paints and Coatings: VOC not more than 150 g/L.
3. Anti-Corrosive Coatings: VOC not more than 250 g/L.
4. Varnishes and Sanding Sealers: VOC not more than 350 g/L.
5. Stains: VOC not more than 250 g/L.
6. Aromatic Compounds: Paints and coatings shall not contain more than 1.0 percent by weight total aromatic compounds (hydrocarbon compounds containing one or more benzene rings).
7. Restricted Components: Paints and coatings shall not contain any of the following:
   a. Acrolein.
   b. Acrylonitrile.
   c. Antimony.
   d. Benzene.

   e. Butyl benzyl phthalate.

   f. Cadmium.

   g. Di (2-ethylhexyl) phthalate.

   h. Di-n-butyl phthalate.

   i. Di-n-octyl phthalate.

   j. 1,2-dichlorobenzene.

   k. Diethyl phthalate.

   l. Dimethyl phthalate.

   m. Ethylbenzene.

   n. Formaldehyde.

   o. Hexavalent chromium.

   p. Isophorone.

   q. Lead.

   r. Mercury.

   s. Methyl ethyl ketone.

   t. Methyl isobutyl ketone.

   u. Methylene chloride.

   v. Naphthalene.

   w. Toluene (methylbenzene).

   x. 1,1,1-trichloroethane.

   y. Vinyl chloride.

 C. Credit EQ 4.4: Do not use composite wood

PART 3 - EXECUTION

 3.01 SITE DISTURBANCE

  A. Credit SS 5.1: Comply with requirements of Division 1 Section "Summary".

 3.02 REFRIGERANT REMOVAL

  A. Prerequisite EA 3.0: Remove CFC-based refrigerants from existing HVAC and refrigeration equipment indicated to remain and replace with refrigerants that are not CFC based. Replace or adjust existing equipment to accommodate new refrigerant as described in Division 15 Sections.

  B. Credit EA 4.0: Remove HCFC-based refrigerants from existing HVAC and refrigeration equipment indicated to remain and replace with refrigerants that are not HCFC based. Replace or adjust equipment to accommodate new refrigerant. Remove clean-agent fire-extinguishing agents that contain HCFCs or halons, and replace with agent that does not contain HCFCs or halons.

   1. Refer to Division 15 Section for additional requirements.

   2. Refer to Division 13 Section "Clean-Agent Extinguishing Systems" for additional requirements.

 3.03 CONSTRUCTION WASTE MANAGEMENT

  A. Credit MR 2.1 and 2.2 Comply with Division 1 Section "Construction Waste Management."

3.04 CONSTRUCTION INDOOR AIR QUALITY MANAGEMENT

A. Credit EQ 3.1: Comply with SMACNA IAQ Guideline for Occupied Buildings under Construction.

1. If Owner authorizes the use of permanent heating, cooling, and ventilating systems during construction period as specified in Division 1 Section "Temporary Facilities and Controls," install filter media having a MERV-8 according to ASHRAE 52.2 at each return-air inlet for the air-handling system used during construction.

2. Replace all air filters immediately prior to occupancy. Replacement air filters shall have a MERV-13 according to ASHRAE 52.2.

B. Credit EQ 3.2:

1. Conduct a two-week building air flush-out after construction ends with new air filters and 100 percent outdoor air. Replace air filters after building air flush-out. Replacement air filters shall have a MERV-13 according to ASHRAE 52.2.

a. Insert specific details of building air flush-out procedure here.

2. Owner will conduct a baseline indoor air quality testing program according to EPA Protocol for Environmental Requirements, Baseline IAQ and Materials, for Research Triangle Park Campus, Section 01445. Payment for these services will be made by Owner.

3. Engage an independent testing and inspecting agency to conduct a baseline indoor air quality testing program according to EPA Protocol for Environmental Requirements, Baseline IAQ and Materials, for Research Triangle Park Campus, Section 01445.

BUILDING CODES AND MUNICIPAL REGULATIONS

The fourth place the contractor will find requirements for the project will be in the applicable building codes. The applicable building code will vary from region to region throughout the United States. Some of these codes have broader applicability like the International Building Code, which according to the International Code Council has been adopted in one form or another by 47 states. Some have more limited applicability like the California State Building Code, the Florida Building Code and the Rhode Island State Building Code. The applicability of these codes is limited to a single state. Although published as a separate state code, these codes are based on the International Building Code. The most limited codes are those which might be applicable to only one city. The Chicago Building Code and the New York City Building Code are two examples of these narrowly applicable codes.

The applicability of these codes is generally established by the municipality in which the code is required. For instance, the "Florida Building Code" has been adopted for use within the entire state of Florida. The code itself references in other national codes such as the National Electric Code, which will govern all electrical work. In fact, most codes defer to the National Electric Code for all electrical requirements and regulations. The Life Safety Code, which governs occupant loads and means of egress is another national code that is commonly referenced into other more local codes.

Together, these codes define the code requirements the contractor must follow in the construction of the building. The building code contains two levels of information. The first level of information contained in the code is the primary level. This primary level of information explicitly defines all of the code requirements for a given type of material or construction. It will be quite detailed and descriptive of the qualities of the work required. The other level of code information is the secondary level. The secondary level of code information is not explicitly described within the code itself. The details regarding this code requirement is found in other codes or standards which are referenced into the code. The following is an example from the Florida Building Code of how the American Concrete Institute Standard 318 is referenced into the code.

FBC 1905.8 Concrete Mixing. Mixing of concrete shall be performed in accordance with ACI 318, Section 5.8.

When this occurs, the contractor must obtain and review the referenced standard in order to understand what compliance will entail. If the contractor fails to review the referenced standard, then there will be a high likelihood of the contractor being found to be in noncompliance.

As mentioned in Chapter 2, in March of 2010, the International Code Council announced the first model building code for green buildings. Other regional and local building codes are more than likely going to be modified to incorporate green building standards. Contractors must be careful to review new editions of their applicable building code for these new green building requirements.

OTHER STANDARDS OR REQUIREMENTS

Other standards or requirements would include standards such as the LEED standards. While these will be referenced in the contract, general conditions or specifications, the details therein are only found in the standard itself. It will be the contractor's responsibility to thoroughly understand the details of both how the construction-related credits can be met. In addition, the contractor must fully understand the LEED documentation requirements for each of those credits. The following is a sample of the LEED 2009 New Construction and Major Renovation Project Scorecard (see Figure 4-12).

Figure 4-12
LEED scorecard.

LEED 2009 for New Construction and Major Renovations

Project Checklist

Project Name

Date

Sustainable Sites — Possible Points: 26

				Credit	Description	Points
Y	?	N				
Y				Prereq 1	Construction Activity Pollution Prevention	
				Credit 1	Site Selection	1
				Credit 2	Development Density and Community Connectivity	5
				Credit 3	Brownfield Redevelopment	1
				Credit 4.1	Alternative Transportation—Public Transportation Access	6
				Credit 4.2	Alternative Transportation—Bicycle Storage and Changing Rooms	1
				Credit 4.3	Alternative Transportation—Low-Emitting and Fuel-Efficient Vehicles	3
				Credit 4.4	Alternative Transportation—Parking Capacity	2
				Credit 5.1	Site Development—Protect or Restore Habitat	1
				Credit 5.2	Site Development—Maximize Open Space	1
				Credit 6.1	Stormwater Design—Quantity Control	1
				Credit 6.2	Stormwater Design—Quality Control	1
				Credit 7.1	Heat Island Effect—Non-roof	1
				Credit 7.2	Heat Island Effect—Roof	1
				Credit 8	Light Pollution Reduction	1

Water Efficiency — Possible Points: 10

	Credit	Description	Points
Y	Prereq 1	Water Use Reduction—20% Reduction	
	Credit 1	Water Efficient Landscaping	2 to 4
	Credit 2	Innovative Wastewater Technologies	2
	Credit 3	Water Use Reduction	2 to 4

Energy and Atmosphere — Possible Points: 35

	Credit	Description	Points
Y	Prereq 1	Fundamental Commissioning of Building Energy Systems	
Y	Prereq 2	Minimum Energy Performance	
Y	Prereq 3	Fundamental Refrigerant Management	
	Credit 1	Optimize Energy Performance	1 to 19
	Credit 2	On-Site Renewable Energy	1 to 7
	Credit 3	Enhanced Commissioning	2
	Credit 4	Enhanced Refrigerant Management	2
	Credit 5	Measurement and Verification	3
	Credit 6	Green Power	2

Materials and Resources — Possible Points: 14

	Credit	Description	Points
Y	Prereq 1	Storage and Collection of Recyclables	
	Credit 1.1	Building Reuse—Maintain Existing Walls, Floors, and Roof	1 to 3
	Credit 1.2	Building Reuse—Maintain 50% of Interior Non-Structural Elements	1
	Credit 2	Construction Waste Management	1 to 2
	Credit 3	Materials Reuse	1 to 2

Materials and Resources, Continued

			Credit	Description	Points
Y	?	N			
			Credit 4	Recycled Content	1 to 2
			Credit 5	Regional Materials	1 to 2
			Credit 6	Rapidly Renewable Materials	1
			Credit 7	Certified Wood	1

Indoor Environmental Quality — Possible Points: 15

			Credit	Description	Points
Y	?	N			
Y			Prereq 1	Minimum Indoor Air Quality Performance	
Y			Prereq 2	Environmental Tobacco Smoke (ETS) Control	
			Credit 1	Outdoor Air Delivery Monitoring	1
			Credit 2	Increased Ventilation	1
			Credit 3.1	Construction IAQ Management Plan—During Construction	1
			Credit 3.2	Construction IAQ Management Plan—Before Occupancy	1
			Credit 4.1	Low-Emitting Materials—Adhesives and Sealants	1
			Credit 4.2	Low-Emitting Materials—Paints and Coatings	1
			Credit 4.3	Low-Emitting Materials—Flooring Systems	1
			Credit 4.4	Low-Emitting Materials—Composite Wood and Agrifiber Products	1
			Credit 5	Indoor Chemical and Pollutant Source Control	1
			Credit 6.1	Controllability of Systems—Lighting	1
			Credit 6.2	Controllability of Systems—Thermal Comfort	1
			Credit 7.1	Thermal Comfort—Design	1
			Credit 7.2	Thermal Comfort—Verification	1
			Credit 8.1	Daylight and Views—Daylight	1
			Credit 8.2	Daylight and Views—Views	1

Innovation and Design Process — Possible Points: 6

Credit	Description	Points
Credit 1.1	Innovation in Design: Specific Title	1
Credit 1.2	Innovation in Design: Specific Title	1
Credit 1.3	Innovation in Design: Specific Title	1
Credit 1.4	Innovation in Design: Specific Title	1
Credit 1.5	Innovation in Design: Specific Title	1
Credit 2	LEED Accredited Professional	1

Regional Priority Credits — Possible Points: 4

Credit	Description	Points
Credit 1.1	Regional Priority: Specific Credit	1
Credit 1.2	Regional Priority: Specific Credit	1
Credit 1.3	Regional Priority: Specific Credit	1
Credit 1.4	Regional Priority: Specific Credit	1

Total — Possible Points: 110

Certified 40 to 49 points   Silver 50 to 59 points   Gold 60 to 79 points   Platinum 80 to 110

RISK MANAGEMENT THROUGH INSURANCE

As the green building movement gains momentum, many owners, architects, and contractors express enthusiasm about sustainable building. For the construction professionals, this enthusiasm is tempered as they weigh the risks involved. As mentioned earlier in this chapter, there are many risks in LEED projects above those commonly identified in traditional construction projects. Marsh, Inc., the world's largest insurance broker has recently published a report entitled Green Building-Assessing the Risks—Feedback from the Construction Industry. The report was based on feedback from 55 design and construction executives involved in the Green Construction movement. In that report, the construction professionals were most concerned about the potential financial risks associated with green construction. In addition, the executives were concerned about being able to maintain the tight schedules required for most projects while still complying with LEED requirements. The report goes on to list the top three risks associated with sustainable construction. They are:

1. *Performance Risks.* This risk involves the ability of products, systems, and equipment to perform at the levels required by the sustainability standards.
2. *Subcontractor Risks.* This risk centers on the selected subcontractors having the knowledge and ability to perform to the requirements for sustainability, stated in the construction documents.
3. *Regulatory Exposure.* This concern involves the uncertainty as to how the regulatory environment, that is, building codes and standards, might evolve with respect to building sustainability. The evolution of building codes and standards to mandate sustainability requirements could open the door to lawsuits that could involve punitive damages. To obtain additional protection, building owners might seek additional long-term warranties from the contractors.

Contractors face certain specific risks when working on a LEED project. By entering into the construction agreement to construct a LEED building, there will be either explicit or implied language requiring the contractor to comply with the construction-specific LEED requirements. They will also generally be required to complete construction within a set amount of time. Constructing a LEED certified building requires considerably more paperwork and documentation than constructing a traditional building. This extensive documentation during the construction process can sometimes result in delays. This can create a liability for the contractor. Several attorneys are questioning the liability contractors might have when specifying "green" products and equipment that are not fully tested. If these products and equipment do not perform to the anticipated standard, they could be considered defects in the construction and thus open the contractor up to liability.

Unfortunately, these risks are new to the insurance industry as well. All contractors carry construction liability insurance. However, according to

Howard Carsman and Janet Kim Lin, LEED accredited construction attorneys, construction liability insurance covers damage to persons or property. Damages that might accrue from a failure to meet the requirement for a certain level of LEED certification does not fall into the commonly understood definition of person or property damage and might not be covered by the contractors liability insurance.[5]

Joseph Fobert, senior vice president and practice leader for the real estate solutions group of Chartis Insurance in Tampa, FL, believes that traditional insurance policies are not adequate to cover contractors working on LEED projects.

> The typical coverage afforded under a CGL policy is for bodily injury and property damage liability, and advertising or personal injury, he noted. Breach of warranty and breach of contract are exclusions, he added, explaining that when a third party alleges a building didn't live up to its green-building hype, insured real estate owners or property managers might not find coverage under a CGL policy for the costs of responding to such charges. The exposures to these individuals are predominately economic and reputational[6]

To protect themselves, several insurance companies are offering a Green Reputation CGL endorsement to the construction policy. However, although this endorsement will provide coverage for defending against the damages reputation claim, it will not provide coverage for the actual cost of correcting the problem. This would require yet another endorsement. It is highly suggested that the contractor meet and thoroughly review the risks of undertaking a LEED project with his or her insurance carrier prior to entering into a LEED building contract.

The insurance industry, as a whole, is evolving with the construction industry as both industries become more educated as to the challenges and risks of green construction. Some construction risk insurance companies are rising to the challenges of insuring contractors working on green projects. Newly developed insurance products offer the following:

- Coverage or fungi, mold including *Legionella pneumophila*
- Nonowned waste disposal sites option
- High limits up to $50 million
- Premium discounts for LEED certified projects
- Long-term policy limits
- Reputational damage endorsements
- Endorsements to cover the cost of LEED compliance corrections

According to All Business a D&B Company, more than 20 insurance carries offer some type of Green-related coverage. These companies include the following:

- Fireman's Fund, http://www.firemansfund.com
- Chubb, http://www.chubb.com
- AIU/Lexington,

- Liberty Mutual, http:// www.libertymutual.com
- Affiliated FM, http://www.affiliatedfm.com
- Philadelphia, http://www.phly.com
- Zurich, http://www.zurichna.com
- CNA, http://www.cna.com

All Business, http://www.allbusiness.com/insurance/insurance-policies-claims-insurance

CONSTRUCTION ACTIVITY POLLUTION PREVENTION

BACKGROUND

Water Pollution by Construction

One of the largest types of pollution created by the construction process is water pollution. This pollution results from water, generally rainwater, washing across construction sites. This storm water is created by either rain or melted snow flowing across impervious ground surfaces.

In the construction of a building, one of the first steps in the construction process is the clearing of the site. During this operation, all trees, bushes, grass, and other vegetation within the area of construction are removed. Once this vegetation is removed, the bare ground is now vulnerable to erosion. Erosion can be defined as the wearing away of the surface of the land by natural forces such as wind and water. Except in very dry climates, water flow is the primary cause of erosion on construction sites. As the water washes across a construction site, it picks up both natural materials such as sand and silt and organic and inorganic pollutants. These organic and inorganic pollutants include heavy equipment fuel, oils, pesticides, and chemicals from other building materials. The amount of these chemical pollutants is generally small when compared with the natural pollutants, that is, sand and silt, generated by the erosion process.

This erosion process is influenced by several factors including the local climate, topography, composition of the soil, and the vegetative cover:

- *Climate:* The local climate factor includes the frequency of the rainfall, that is, how often it is expected to rain, the intensity of the rainfall, which is how many inches of rain can be expected within a given amount of time, and the duration of the rainfall, which is a measure of how long it is expected to rain. All of this historic rainfall information is available from the National Weather Service at www.weather.gov.

  The sample table below compares monthly rainfall amounts from 2010 to the 30-y. averages for a sample city (see Figure 5-1).

A contractor when preparing for a project in Atlanta can ascertain the most recent monthly rainfall for that city. In addition, the contractor can compare the most recent figures with a 30-y. average. By doing this comparison, the contractor can more accurately predict the potential for rainfall and the severity during the construction process. For instance, in the case of this sample city, the months of April, May, June, August, and October generally have the least rainfall. This

Monthly Average Rainfall in Inches

		Jan	Feb	Mar	Apr	May	June	July	Aug	Sep	Oct	Nov	Dec	Total	Difference from 30 year average
Sample City	2010	3.95	2.63	1.31	1.79	2.05	3.66	1.85	3.48	2.92	2.47	0.96	4.78	31.85	−18.35
1971– 2000	30 yr avg	5.03	4.68	5.38	3.62	3.95	3.63	5.12	3.67	4.09	3.11	4.10	3.82	50.20	

COURTESY OF NATIONAL WEATHER SERVICE, WWW.WEATHER.GOV

Figure 5-1

The sample table compares monthly rainfall amounts from 2010 to the 30-y. averages for a sample city.

information can be helpful in scheduling the construction process in a way to protect against excessive soil erosion due to rainwater runoff.

- *Topography:* The topography factor deals with the slope of both the site and the surrounding areas. The steeper the slope, the faster the storm water runoff will be. The contractor must consider both the macro and micro topography when planning the project. At a closer view, the project site might be relatively flat, so no significant water issues might be predicted. However, when looking from a macro standpoint, the flat site might be at the bottom of a hill. Now the potential for water flooding issues is almost guaranteed.
- *Soils:* The composition of the soils has a significant effect on the amount of storm water runoff that will be created from a given storm. Every type of soil has a determined percolation rate. This "perc" rate is an index of how much water will be absorbed through the soil over a given time. Soils with higher percolation rates will absorb storm water faster than those soils with a lower percolation rate. This percolation rate is measured in a depth of water absorbed by the soil in 1 h. For instance, soils made up of sandy loam will absorb 20–30 mm of water per hour, whereas a soil composed mostly of clay will only absorb 1–5 mm per hour (see Figure 5-2).
- *Vegetation:* The amount of vegetation on the surface of the ground has another significant influence on the amount of erosion resulting from storm water runoff. Ground surfaces that contain vegetation are much less apt to encounter erosion than those that are bare or sparsely vegetated.

Figure 5-2

Percolation rates for soils.

Percolation Rates for Soils	
Soil Type	Infiltration Rate in mm/h
Sand	Less than 30
Sandy Loam	20–30
Loam	10–20
Clay Loam	5–10
Clay	1–5

COURTESY OF THE FOOD AND AGRICULTURE ORGANIZATION OF THE UNITED NATIONS

It has been estimated that runoff from a cleared, unstabilized construction site can result in the loss of approximately 35–45 tons of sediment per acre each year.[1]

The Environmental Protection Agency identifies five major types of erosion encountered on a construction site. They are as follows:

1. Raindrop erosion, which is defined as the dislodging of soil particles by raindrops.
2. Sheet erosion, which is defined as the uniform removal of soil by the action of a sheet flow of water. The unique characteristic of this type of erosion is that it leaves no identifiable channels or marks on the ground.
3. Rill erosion, which is a type of erosion that results from concentrated runoff. This type of erosion creates many small channels in the ground within which the water flows.
4. Gully erosion, which is similar to rill erosion except that the flow is much more concentrated. The result is the creation of deep water channels in the ground.
5. Streambank erosion, which is a type of erosion that erodes unstable banks of streams (see Figures 5-3 and 5-4).

Although the loss of earth on the site can be a considerable problem, where that sand and silt is finally deposited presents a much greater problem. Suspended solids, that is, silt is one of the greatest pollutants of our nation's waterways. These suspended solids can be very injurious to fish and other aquatic life living in the stream. These solids increase the turbidity of the water. This increased turbidity reduces the ability of sunlight to penetrate into the water, which reduces the photosynthesis of the aquatic plants. This in turn reduces the amount of oxygen in the

Figure 5-3

Rill erosion.

Figure 5-4

Gully erosion.

PHOTO COURTESY OF USDA NATURAL RESOURCES CONSERVATION SERVICE

water. The lower oxygen level results in the loss of life to fish and other aquatic species in the water body. The loss of these fish can create a ripple effect throughout the local biosphere. Birds and other animals that feed on these fish will also be affected.

Clean Water Act

In response to a greater understanding of the long-term effects of erosion, in 1972, the United States Congress passed the Clean Water act (CWA). This act established the basic legal structure for regulating discharges of pollutants into the waterways within the United States. The responsibility for enforcing the act went to the EPA. Toward this goal of regulation and enforcement, the EPA established standards for any wastewater deposited into our nation's waterways (see Figure 5-5):

- All interstate waters
- Intrastate waters used in interstate and/or foreign commerce
- Tributaries of the above
- Territorial seas at cyclical high-tide mark
- Wetlands adjacent to all of the above

In addition, the EPA made it illegal to discharge any pollutants from a point source into any navigable waterway unless a special permit is first obtained.

Figure 5-5

Clean water stream.

COURTESY OF FIR0002/WIKIPEDIA

A point source is defined as discharge from a man-made source such as a ditch or pipe. Residential septic tanks are not included and do not require a permit. The CWA has four different levels of source pollution situations that require permits. The following is a list of those pollution situations regulated by the CWA (see Figure 5-6):

- *Point Sources of Pollution*
  The National Pollutant Discharge Elimination System (NPDES) controls and regulates point sources of pollution discharging into a surface waterbody such as a lake or stream.
- *Nonpoint Pollution Sources*
  The act regulates a multitude of nonpoint sources of pollution including but not limited to the farming and forestry industries.
- *Placement of Dredged Materials*
  The act controls and regulates the placement of dredged materials into waters of the United States. The act also governs dredged materials deposited into our nation's wetlands.
- *Pollution of Waterways*
  The act requires that Federal agencies obtain a certification for a state or territory prior to issuing permits for operations that might result in an increased load of the pollutants deposited in a waterway.

Figure 5-6

Polluted stream.

COURTESY OF DAVID A. VILLA

INTENT OF THE LEED REQUIREMENT

The need to control pollution during construction activities is viewed as so important by USGBC that it has become a mandatory requirement on all LEED projects.
The requirement for construction activity pollution prevention falls under the
Sustainable Site Section and is listed as Prerequisite 1 (SS Pre-1). The intent of
this LEED requirement is to reduce air and water pollution from all construction
activities on LEED projects. This requirement is a prerequisite in the LEED standards and its compliance is mandatory on all LEED projects. It is expected that
LEED buildings are not only more energy efficient to operate but that their construction process will have a reduced impact on the environment when compared
with conventional "non-LEED" construction.

LEED REQUIREMENTS FOR CONSTRUCTION ACTIVITY POLLUTION PREVENTION

The primary demand of this prerequisite is the development of a plan to control erosion. The plan must be completed and enacted prior to the commencement of construction. The plan must meet either the requirements of the 2003
EPA Construction General Permit or the requirements of the local jurisdiction or
applicable codes, whichever is more stringent.

All construction activities that will disturb one or more acres are regulated by
the EPA. This prerequisite essentially applies the NPDES to all LEED projects,

regardless of size. On March 10, 2003, the new NPDES regulations came into effect. These new regulations extend the previous coverage of the regulation to construction sites that disturb 1–5 acres. This would include smaller building sites that are part of a larger development plan. The previous regulations only regulated sites 5 acres or larger. Details on the EPA construction general permit can be found at http://cfpub.epa.gov/npdes/stormwater/cgp.cfm.

TAKE NOTE

During construction prevent
- Soil loss
- Sediment deposits in waterways
- Air pollution

The Pollution Prevention Plan

The Pollution Prevention Plan, which is a requirement on all LEED projects regardless of size, must include measures to prevent:

- Soil loss, including topsoil, from rain and wind erosion during construction
- Sediments from entering local sewers or other waterways
- Air pollution in the form of dust and other particles

The EPA has published a Pollution Plan Template for use on every project in all states. The contractor is able to customize this form to reflect the conditions of the EPA General Permit and the specific conditions of the site. The Pollution Plan consists of eight major sections; they are as follows:

Section 1: Site Evaluation, assessment, and planning

1.1 Project/Site Information
1.2 Contact Information/Responsible Parties
1.3 Nature and Sequence of Construction Activity
1.4 Soils, Slopes, Vegetation, and Current Drainage Patterns
1.5 Construction Site Estimates
1.6 Receiving Waters
1.7 Site Features and Sensitive Areas to be Protected
1.8 Potential Sources of Pollution
1.9 Endangered Species Certification

Section 2: Erosion and sediment control best management practices

Section 3: Good housekeeping best management practices

Section 4: Selecting post-construction best management practices

Section 5: Inspections

Section 6: Record keeping and training

Section 7: Final stabilization

Section 8: Certification and notification

STRATEGIES FOR SUCCESS

Strategy 1: The Soils Report

One of the first steps in the compliance of this LEED prerequisite is to gain an understanding of the soils involved. If the soils information is not a part of the construction documents developed by the design team, a soils testing service should be retained to ascertain the type and properties of the soils on the site. The result of this testing will be a report outlining the properties of the soils including the stability of the soil and the ability of the soil to percolate water. There are two major types of soils reports available to the contractor. The first is a soil boring test and the second is a percolation or exfiltration test.

- *Soil boring test*

The soil boring test, which is also referred to as a standard penetration test, consists of a hollow steel tube called a split spoon, which is driven into the ground. The tube is driven by a drop hammer, which is capable of exerting a constant force through impacting a series of blows onto the end of the tube. Figure 5-5 illustrates how a standard penetration test is undertaken.

During the test, the number of blows needed to drive the tube every 6 in. is recorded. This number of blows is an indication of the density of the soil and thus its ability to bear loads. This information is recorded on a soil boring log. Since the tube is hollow, in addition to being able to measure the density of the soil, the tube itself collects a sampling of the soil. Once the tube is returned to the lab, it is laid out on a long table and the tube is split open exposing a core of soil. The soil core is then examined, and the different types of soil are identified and recorded in the log. The following is an example of a soil boring log (see Figure 5-7).

The above soil boring log indicates that a level of soil extending 7 ft. below the grass and topsoil consists of loose sand. This upper layer is followed by another 7-ft. layer of silt. The third layer consists of approximately 7 ft. of denser sand. The increased density of the soil is evidenced by the increased number of impacts required to penetrate 1 ft. of depth into the ground. To the far right, there appears a note that indicates the groundwater is approximately 15.5 ft. below the upper surface of the ground. The table continues to depict both the type of soil and the relative density as measured in the blows per foot for the remaining boring until the sampler hits solid rock at 64 ft. There are a total of nine different layers encountered from the grass at the surface to the solid rock below. The information derived from this test will allow the contractor to anticipate how the subsurface soils will behave during the construction process. For instance, very loose soils might require the forming of the building foundations. Foundations constructed in firmer soils might not

Figure 5-7

Sample soil boring log.

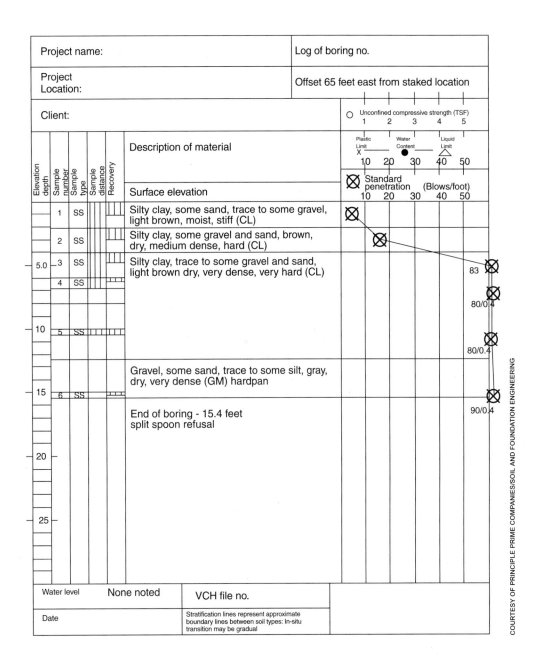

require forms. The requirement for forms will add a significant amount to the building construction cost.

- *Soil percolation test*

In addition to understanding the composition of the soils, it is important for the contractor to understand how water percolates through the soil. The soil's ability to allow storm water to pass through it will drastically affect the potential for soil erosion. For instance, a soil like sand has the ability to allow storm water to pass through it very rapidly. Because of this rapid penetration of the water through the soil, there is a reduced amount of runoff and thus a lower probability for erosion. Assume the soil was composed of clays or other soils that do not allow water to penetrate as fast. In this case, there will be much more surface water running off the site and this will create a greater potential for soil erosion (see Figure 5-8).

REPORT OF DETERMINATION OF HYDRAULIC CONDUCTIVITY USING USUAL CONDITION TEST DATA

CLIENT : THE CITY OF SUNSHINE

      Attn: Mr. John Smith

      Project Manager

      1234 Sunset Road

      Sunshine, FL 32123

PROJECT: SUNSHINE PARK

      PROPOSED NEW CONSTRUCTION

      8000 Sunset View Drive, Sunshine, Florida

      Hydraulic Conductivity ("K" Factor)

LOCATION: As noted on attached Test Location Plan

In accordance with your request and authorization, a representative of Baier Consulting Engineers, Inc. conducted one (1) Usual Condition Exfiltration Test at the above referenced project. We are herewith providing you with Hydraulic Conductivity calculations in connection with this test.

The test was performed in accordance with the City of Sunshine Water Management Authority procedures; specifically, the Usual-Open Hole Test.

Test results, including hydraulic conductivity ("K" Factor) computations, are shown on the enclosed Data Sheet.

Should you have any questions, please do not hesitate to contact us.

Respectfully submitted,

Ronald A. Baier, P.E. # 16211
Consulting Engineer

REPORT OF DETERMINATION OF HYDRAULIC CONDUCTIVITY

PROJECT: CITY OF SUNSHINE-SUNSHINE PARK

TEST NO: HC-1 of 1

$$K = \frac{4Q}{\pi d (2 H_2^2 + 4 H_2 Ds + H_2 d)}$$

WHERE:

K = HYDRAULIC CONDUCTIVITY (cfs/ft2 − ft. Head)

Q = "Stabilized" FLOW (cfs) = 0.00220572 [0.99 GPM]

d = DIAMETER OF TEST HOLE (ft.) = 0.50

H_2 = DEPTH OF WATER TABLE (ft.) = 2.167

D_s = SATURATED HOLE DEPTH (ft.) = 7.833

K = 0.0000717 or K = 7.17×10^{-5}

The result of this test is a "K" value which is a measure of how much water is absorbed into the

Figure 5-8
Soil percolation test.

Figure 5-9

Sample sedimentation control plan.

Control Strategy	Description
Vehicle Tracking	Approximately 8,000 SF of lawn and topsoil will be removed and replaced by a free draining gravel material to allow truck access with minimal soil displacement.
Silt Fencing	Silt fencing with straw bales will be installed along the north, east and west elevations. The south elevation will have a silt fence barrier without straw bales.
Sediment Basin	The site has two existing catch basins. One on the west and one on the north plus nine smaller catch basins on the south elevation. Each catch basin will have sediment traps. One existing catch basin located within the site perimeter fence will have a sediment trap with a straw bale barrier.
Inspections	Silt fencing and sediment traps will be inspected and maintained on a weekly basis by the general contractor. In the event of significant rainfall, controls will be inspected at the end of the workday or the following morning.
Permanent Seeding and Planting	Undisturbed site areas containing existing landscaping and trees will be protected from truck and vehicle traffic. Upon completion disturbed areas of the site will be immediately seeded and planted with permanent vegetation.

Strategy 2: The Sedimentation Control Plan

As mentioned in the previous section, the development of a comprehensive plan to control and mitigate the damage caused by sedimentation is a critical part of this process. The following is an example taken from the LEED Reference Guide, which illustrates such a plan (see Figure 5-9).

Strategy 3: Storm Water Runoff Remediation

During the construction process, there are several things a contractor can do to mitigate the damage caused by storm water. One of the first and the most effective methods of mitigating the damage caused by soil erosion is the installation of silt barriers. These barriers can be easily installed along the perimeter of the construction site. Their purpose is to control and filter the storm water runoff so as to eliminate or substantially reduce the amount of sediment in the runoff water. An important note to remember is that the erosion control device should be installed within 24 h. of the initial soil disturbance. These installations can take any of the following forms.

- *Silt fences*

  A silt fence is a temporary barrier installed along the perimeter of a construction site. It consists of a geotextile fabric supported by posts driven into the ground. This geotextile fabric is a woven fabric that performs three basic functions (see Figures 5-10 through 5-12).

Figure 5-10

Silt fence erosion protection.

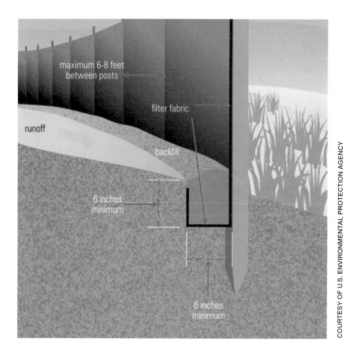

maximum 6-8 feet between posts

filter fabric

runoff

backfill

6 inches minimum

6 inches minimum

COURTESY OF U.S. ENVIRONMENTAL PROTECTION AGENCY

Figure 5-11

Silt fence.

COURTESY OF U.S. ENVIRONMENTAL PROTECTION AGENCY

Figure 5-12

Stormwater inlet
silt protection.

COURTESY OF USDA/NRCS

1. *Filtration:* The geotextile fabric's woven design is fine enough to retain even the finest sands and silt while still allowing storm water runoff to pass through it. This silt filtration eliminates or significantly lessens the potential for pollution of adjacent sites or waterways.
2. *Drainage:* The design of the silt fence installation serves to control the direction of the water runoff and flow of water. The storm water can be retained and redirected to a designed collection basin.
3. *Separation:* The installation of the silt fence serves to separate different type of soils. The fence would not allow one soil to cross contaminate the adjacent type of soil.

- *Straw bales*
  The use of straw bales can also perform a similar function to a geotextile silt fence. A straw bale is a bundle of straw that has been bound tightly with either twine or wire. These bales can be either cylindrical or rectangular depending on the balking equipment used to form the bales. This somewhat less sophisticated method uses a linear arrangement of straw bales. The bales are recessed into the ground and staked into place; once installed, they will serve to filter and direct storm water runoff. Many state and local governments allow straw bale silt fences as a sole barrier, but the EPA does not recommend the use of straw bale when other geotextile methods are available. Occasionally, straw bale silt fences are used as an effective secondary barrier in conjunction with a conventional silt fence. The following photo illustrates the use of a straw bale barrier in conjunction with two lines of geotextile silt fences (see Figure 5-13).
- *Temporary seeding*
  Another method of mitigating the potential for pollution from storm water runoff is the use of temporary seeding. This involves the use of fast-growing grasses and ground cover including grass, legume, and grain

Figure 5-13

Secondary straw bale erosion control.

seed. Like all mitigation methods, the seed must be planted immediately after the initial soil disturbance. If the ground has been compacted from the movement of heavy construction equipment, steps should be taken to loosen the soil to a depth of between 6 and 8 in. prior to planting. Keep in mind that the grasses and groundcover resulting from this seeding will only be effective in erosion control if it is protected from further construction disturbance. Seeding should be plated at the following rates (see Figures 5-14 and 5-15).[2]

Figure 5-14

Temporary seeding rates.

	Temporary Seeding Species and Rates		
Species	Lbs Seed per Acre	Lbs of Seed per 1000 sf	Comments
Oats	80	2	Height up ton 2 ft. Not cold tolerant.
Rye	90–120	2	Height up to 3 ft. Cold and low pH tolerant.
Wheat	90–120	2	Height up to 3 ft. Cold and low pH tolerant.
Milletts	45–60	1–1.5	Height up to 5 ft. Aggressive growth.
Annual Ryegrass	75	2	Height up to 16 in. Not heat tolerant
Annual Lespedeza	15	0.5	Warm season tolerant. Tolerates low pH.
Tall Fescue	45	1	

Figure 5-15

Temporary seeding for erosion control.

- *Earth dikes*

  An earth dike is another type of erosion control. It consists of a temporary linear mound or ridge of earth constructed in combination with a shallow trough or drainage swale. The ridge serves to control and direct the storm water runoff to the drainage swale, which redirects the flow another direction thus avoiding a potential pollution situation. Earth dikes do have limitations. They cannot be constructed of soils that are easily eroded and they are not considered suitable sediment trapping devices. Once constructed, they must be stabilized immediately, and on smaller sites, they can often become barriers to construction equipment. The following chart illustrates criteria for the construction of erosion control dikes (see Figures 5-16 through 5-18).

Figure 5-16

Earth dike design criteria.

Criteria	Drainage Area < 5 Acres	Drainage Area between 5 and 10 Acres
Dike Height	18 in.	36 in.
Dike Width	24 in.	36 in.
Flow Channel Width	48 in.	72 in.
Flow Channel Depth	8 in.	15 in.
Side Slopes	2:1 or flatter	2:1 or flatter
Grade	0.5%–20%	0.5%–20%

Figure 5-17

Cross section drawing through earth dike.

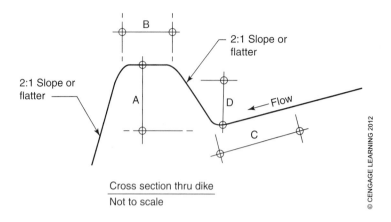

2:1 Slope or flatter

2:1 Slope or flatter

Flow

Cross section thru dike
Not to scale

© CENGAGE LEARNING 2012

Figure 5-18

Earth dikes for erosion control.

PHOTO BY BRUCE YOUNG, COURTESY OF ST MARY SOIL CONSERVATION DISTRICT

Mulching

Mulching is another common method of preventing soil erosion. The mulch is commonly made of organic materials such as straw or wood fiber. Wood mulch can be applied by hand or with the aid of a wood chipper. It can consist of both soft and hard wood pieces. Care must be taken when dealing with wood chip mulch to assure that the wood does not contain any man-made waferboard or particle board and that has not been treated with any type of chemical preservative such as chromated copper arsenic CCA or penta-treatment. These have harmful chemicals that will leach into the ground and pollute the groundwater supply. Straw mulch is generally applied in blankets. These blankets of straw are secured by nets to the unstable earth. In either method, the mulch serves to protect the unstable soils from storm water erosion (see Figure 5-19).

LEED ONLINE DOCUMENTATION

There is a LEED Online documentation form for Site Selection Prerequisite 1, which must be completed by the contractor or their representative. The first step in the documentation process is to select the compliance option the contractor is going to comply with. There are two primary options. The contractor can choose to comply with the 2003 EPA General Construction Permit or they can choose to comply with the applicable local standards and codes. To document compliance, the contractor must either upload the Erosion and Sedimentation Control Plan or they can check off "streamline" the Civil Engineer exemption. To do this, the submitter must be a civil engineer, and therefore no ESC plan is required. Assuming the contractor is not a civil engineer, the full documentation option must be selected. Following the check off, the entire ESC plan must be uploaded to the Web site.

To assure compliance, the contractor must select one of the three verification options. The first involves periodic inspections with documentation providing proof that the ESC plan was carried out properly. The second option allows the contractor to submit photos indicating compliance with the ESC plan. The third option allows the contractor to provide a detailed narrative discussing how compliance with the ESC plan was achieved.

Figure 5-19

Erosion control mulching.

COURTESY OF MAIN STREET MATERIALS

REFERENCES AND SOURCES

- EPA Construction General Permit at http://cfpub.epa.gov/npdes/stormwater/cgp.cfm
- LEED Registered Project Tools at www.usgbc.org/projecttools
- U.S. EPA, Erosion and Sedimentation Control Model Ordinances at www.epa.gov/owow/nps/ordinance/erosion.htm
- United States Department of Agriculture, Natural Resource Conservation Service, NRCS, at http://www.nrcs.usda.gov
- CPESC, Inc. at http://www.cpesc.net. This is a site that contains listings of certified erosion and sedimentation control professionals by state.
- California BMP Storm water Handbook at www.cabmphandbooks.com
- Erosion Control Technology Council at http://www.ectc.org
- International Erosion Control Association at http://www.ieca.org

CONSTRUCTION WASTE MANAGEMENT

6

BACKGROUND

The EPA estimates that there are approximately 1900 landfills operating in the United States. These vary in size depending on the size of the communities they serve. The largest landfill in the United States is called the Freshkills Landfill and is located on Staten Island in New York City. It is so large that it is one of the only two man-made objects that can be seen from space. The other man-made object visible from space is the Great Wall of China.

The types and amounts of material entering a landfill range from household trash and garbage to construction waste and debris (see Figure 6-1).

CONSTRUCTION AND DEMOLITION WASTE

In 1998, the U.S. Environmental Protection Agency undertook a study of the types and amounts of construction-related debris entering municipal landfills. In its report, the EPA generally defined the source of this debris in its broadest terms. Their figures generally include debris from new building construction, renovation, and demolition. The figures also often included C&D debris generated from infrastructure construction activities including the C&D of highways and bridges. In the 1998 report, the EPA included only building C&D debris. They estimated that over 136 million tons of C&D debris are generated annually by building C&D operations The estimate indicates that C&D waste accounts for up to 40% of the national solid waste stream. In a 2003 EPA update, this figure had increased to 164 million tons. This C&D waste can be roughly attributed to the following three sources:

- Construction waste 9%
- Renovation waste 38%
- Demolition waste 53%

Clearly, the volume of waste generated by the construction industry has the potential for creating a significant environmental impact on the United States and perhaps globally. The EPA estimates that currently only about 20% of all C&D waste is diverted from the landfill. The construction industry, beginning with individual contractors, should strive to significantly increase the volume of C&D waste that is diverted.

The EPA report identified concrete as the single largest component of C&D debris. Concrete is followed by wood and drywall as the next two largest C&D waste

Figure 6-1

Waste deposited in landfills 2007.

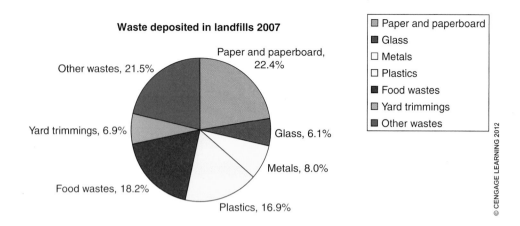

Waste deposited in landfills 2007

- Paper and paperboard, 22.4%
- Glass, 6.1%
- Metals, 8.0%
- Plastics, 16.9%
- Food wastes, 18.2%
- Yard trimmings, 6.9%
- Other wastes, 21.5%

Legend:
- Paper and paperboard
- Glass
- Metals
- Plastics
- Food wastes
- Yard trimmings
- Other wastes

© CENGAGE LEARNING 2012

stream components. Together, these three materials comprise 65%–95% of the total volume of C&D debris entering a landfill. In addition to these three primary building materials, C&D waste also includes other materials such as metals, masonry, and cardboard. The following chart illustrates the breakdown of these other C&D debris materials (see Figure 6-2).

It is common knowledge that most of the debris generated by construction actually comes from demolition operations. The EPA study confirmed this and showed that the debris generated by demolition operations far exceeds that generated by actual construction operations. In fact, a large percentage of the C&D debris stream is generated from building demolition. Not surprisingly, the debris generated by commercial C&D operations far exceeds that generated by residential C&D endeavors. In addition, there are differences between the debris generated by residential construction versus nonresidential construction. This difference often varies depending on the material. The volume of concrete debris is generally greater in the nonresidential sector, whereas wood debris is greater in the residential sector.

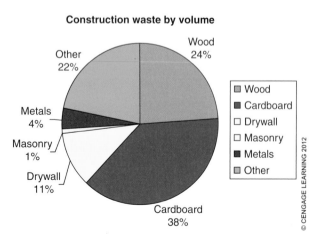

Construction waste by volume

- Wood 24%
- Cardboard 38%
- Drywall 11%
- Masonry 1%
- Metals 4%
- Other 22%

Legend:
- Wood
- Cardboard
- Drywall
- Masonry
- Metals
- Other

© CENGAGE LEARNING 2012

Figure 6-2

Construction waste by volume.

Concrete Debris

For instance, new residential construction generates approximately 0.3 million tons of concrete debris per year. New nonresidential construction contributes another 0.2 million tons annually. Because of the volume of concrete debris generated by residential driveway renovation, residential renovation is said to generate 13 million tons of concrete debris per year. Nonresidential renovation generates approximately 9.8 million tons of concrete debris per year.

The amount of concrete debris generated by demolition is much greater. Residential demolition and commercial demolition generate 6.5 and

29.8 million tons of concrete debris annually, respectively. The sheer size of commercial building demolition as compared with the demolition of residential buildings accounts for the large difference in the amount of concrete debris.

Wood Debris

As stated in Chapter 1, wood is the single largest material used in the residential construction market. This increased use of wood invariably results in an increase of wood in the debris stream. Wood C&D debris range from small pieces generally referred to as "cutoffs" to larger size pieces of wood beams and trusses. In addition, there is a considerable amount of wood-based sheet goods debris that results from both construction and demolition operations. This can be in the form of plywood, OSB, or medium density board (particleboard). According to a report published in BioCycle, residential construction generates approximately 6.6 million tons of wood sheathing waste per year. Residential renovation contributes another 5.2 million tons of wood debris annually. Commercial construction, because of the limited use of wood as a building material, only contributes 4.2 million tons of wood debris per year (see Figure 6-3).

Figure 6-3

Waste wood.

COURTESY OF HOUSTON-GALVESTON AREA COUNCIL

Drywall Debris

Since the 1950s, drywall or gypsum wallboard has become the material of choice for interior wall finishes on both residential and nonresidential projects.

Figure 6-4

Drywall construction waste.

COURTESY OF PHOTOGRAPHER JOSEPH HECKMAN

Figure 6-5

Residential to commercial construction waste comparison.

	Construction Waste Generated Annually			
	Construction		Demolition	
Material	Residential	Nonresidential	Residential	Nonresidential
Concrete	13.3 million tons	10.0 million tons	6.5 million tons	29.8 million tons
Wood	11.8 million tons	4.2 million tons		
Drywall	3.2 million tons	9.5 million tons		9.5 million tons

© CENGAGE LEARNING 2012

This increase in use has made drywall the third largest component in the C&D waste stream. Residential construction accounts for 1.4 million tons of drywall debris added to the waste stream. Due to the large numbers of kitchen and bathroom renovations undertaken in the United States annually, residential renovation adds another 1.8 million tons of drywall debris. Nonresidential construction accounts for 9.5 million tons of drywall debris per year and nonresidential renovations adds another 9.5 million tons per year (see Figures 6-4 and 6-5).

BENEFITS DERIVED FROM WASTE DIVERSION

There are several inherent benefits derived from the diversion of C&D waste from our nation's landfills. These benefits include both environmental and financial benefits. As discussed in Chapter 1, a significant amount of energy goes into making the materials we use to construct our homes and offices, religious facilities, and schools. The amount of energy required to generate new materials from recycled resources is a fraction of that required to develop them from natural materials. This savings in energy can also result in an overall savings in total cost of these materials. In addition, the increased first use of materials is severely taxing our nation's natural resources. By using recycled materials, we can significantly slow down the damage caused to our natural environment.

There are many C&D materials that can be recycled and a multitude of markets currently exist for these materials. The following chart taken from *Construction Business Owner*, June 2007, illustrates some of the potential uses of recycled C&D materials (see Figure 6-6).

Secondly, there is a direct economic benefit to the contractor. As every contractor knows, there is a cost associated with the loading, removal, and disposal of construction debris. Construction debris is commonly collected on site in construction dumpsters. It has been estimated that it takes approximately 2.4 h. per ton of construction debris to load the dumpster. The cost of renting these dumpsters including their removal and dumping can sometimes be a considerable amount. The following sections explain in greater detail some of the cost benefits to the contractor.

Figure 6-6

Uses of recycled C&D materials.

Material	How is it Recycled	Recycling Markets
Concrete	The concrete material is crushed. The reinforcing steel is removed and the remaining material is screened for size.	• Road base • General construction fill • Drainage media • Pavement aggregate
Asphalt Pavement	The asphalt is crushed and then recycled back into asphalt. This can occur on site or at the hot mix asphalt plant.	• Aggregate for new asphalt pavement • Sub base for a new road
Asphalt Shingles	After separation from other materials such as nails and screws the shingles are ground and recycled into hot-mix asphalt.	• Asphalt binder and fine aggregate for hot-mix asphalt.
Wood	After cleaning and removal of nails and screws the wood can be re-milled, chipped, or ground.	• Stock for engineered particleboard • Boiler fuel • Reused as flooring lumber • Mulch and compost • Animal bedding
Drywall	After removal of the paper surfacing the drywall is broken up and ground.	• New gypsum wallboard • Manufacturing of cement • Agricultural applications
Metal	Melted down and reformed	• Metal products
Cardboard	The cardboard is ground up and used in the manufacturing of new cardboard materials.	• Paper products

COURTESY OF U.S. ENVIRONMENTAL PROTECTION AGENCY

TAKE NOTE

Increased profit

Commercial contractors can experience a savings of 30%–90% in their waste disposal costs by recycling.

Residential contractors can directly increase profits on each house built through recycling C&D waste.

Cost of Waste Diversion on Commercial Projects

In all areas of business, the cost of undertaking a new method of working is always compared with the cost of the status quo. In this case, the status quo for C&D waste is to collect the waste from the site and dispose it of in available landfills. If the cost of traditional waste collection and disposal is less than the cost of recycling the waste, there will always be a valid argument against recycling. However, what if recycling C&D waste could actually result in a cost savings to the contractor? This would result in an immediate increase in project profit. The following is an example study undertaken by the Institution Recycling Network. The study compared the cost of traditional waste disposal with the cost of recycling C&D waste in the Boston area. The following chart illustrates the results of this study (see Figure 6-7).

This example clearly illustrates that there is a significant cost savings achieved by recycling versus using traditional C&D waste disposal methods. A 30.8% savings was achieved by recycling commingled wastes rather than using traditional C&D waste disposal methods. The individual material disposal savings range from 36.7% for commercial roofing to 94.8% for metals.

Cost of Waste Diversion on Residential Projects

In 2000, a waste management study was untaken in Alemeda County, California. The study is based on a typical 2000-sq. ft. single-family residence, situated on a generally flat lot. The costs used in the study were based on gate rates published for the area. The results of this study confirm that the recycling of construction waste as opposed to typical disposal in a landfill can result in substantial savings. In fact, this study showed a total savings obtained by recycling of $598.00 or 78% over the cost of typical waste disposal (see Figure 6-8).

	Recycling Cost	Transportation Cost	Total Cost	Total Savings	Percentage of Savings
Doors and Windows	$20.00	$50.00	$70.00	$66.00	48.5%
Gypsum Wallboard	$40.00	$32.00	$72.00	$64.00	47%
Asphalt Shingles	$40.00	$17.00	$57.00	$79.00	58%
Metals	$15.00	$12.00	$27.00	$129.00	94.8%
Glass	$30.00	$21.00	$51.00	$85.00	62.5%
Commercial Roofing	$58.00	$28.00	$86.00	$50.00	36.7%
Clean Wood	$55.00	$29.00	$74.00	$62.00	45.5%
Bathroom Fixtures	$35.00	$16.00	$51.00	$85.00	62.5%
62.5% Concrete, Brick, Block	$10.00	$11.00	$21.00	$115.00	84.5%
Mixed Debris Recycling	$60.00	$34.00	$94.00	$42.00	30.8%
Traditional C&D Disposal	$105.00	$31.00	$136.00		

Figure 6-7

Cost of recycling vs disposal in the Boston area.

© CENGAGE LEARNING 2012

Figure 6-8

Recycling savings on a 2000 SF single family residence.

Material	Tons	Recycling Cost	Recycling Rate	100% Disposal Cost
Wood	3.5	$98.00	80%–100%	$329.00
Drywall	0.73	$0.00	100%	$68.62
Concrete	1.3	$13.65	100%	$122.20
Cardboard	0.28	-$12.60	100%	$26.32
Scrap Metals	0.33	-$12.45	100%	$31.02
Mixed Materials	2	$90.00	60%–85%	$188.00
Total	**8.14**	**$167.60**	**78%–96%**	**$765.16**

COURTESY OF CITY OF OAKLAND, CA

METHODS OF WASTE DIVERSION

There are two primary methods of dealing with C&D waste diversion, they are on-site or source waste separation and off-site or commingled waste separation.

Source Separation

In the on-site method, the contractor is responsible for the waste diversion process. A diversion plan is prepared, separate containers for individual materials to be diverted are positioned at convenient locations on the site. Then individual materials such as concrete, steel, drywall, and so on are collected. Once collected, the material recycler will pick up these materials for processing (see Figure 6-9).

Figure 6-9

Separated
construction waste.

COURTESY OF VT DEC, WASTE MANAGEMENT DIVISION

Figure 6-10

Com-mingled
construction waste.

COURTESY OF HOUSTON-GALVESTON AREA COUNCIL

Commingled Diversion

In the second method of material separation, all C&D materials are collected together in the same container. This is referred to as commingled waste. The recycler then picks up the commingled wastes and separates it off-site. In this method, the contractor is only responsible for collecting the waste. The separation and documentation is the responsibility of the recycling company (see Figure 6-10).

The following chart lists some of the advantages and disadvantages of these two methods of waste diversion (see Figure 6-11).

TAKE NOTE

Source separation versus commingling

Consider commingling recycling. The cost savings is a little less but it is much easier and involves less paperwork.

Recycling Method	Advantages	Disadvantages
Source Separation	• The recycling rate is higher because the recycler is responsible for separating the materials to be recycled.	• There are multiple waste containers on site. Often a separate container will be required for each different material intended to be recycled.
	• The total cost might be lower because the value of the materials to be recycled offsets some of the cost paid by the contractor	• The contractor's employees must spend time separating the selected materials.
	• The contractor can maintain a cleaner site which might result in a safer work environment.	• The contractor is responsible for coordinating his or her on site efforts with several different recycling companies.
		• The contractor will be responsible for documenting compliance with waste diversion recycling goals.
Commingled Recycling	• The contractor has fewer waste containers on the site to coordinate. Often only one or two containers will be required.	• The recycling rated charged by the recycler is generally lower.
	• There is less on site labor cost because the contractor's employees are not responsible for separating the waste materials.	• The overall diversion cost will be higher.
	• The required documentation is prepared by the recycling company so the contractor has lees compliance paperwork to deal with.	

© CENGAGE LEARNING 2012

Figure 6-11

Advantages and disadvantages of C&D waste recycling methods.

INTENT OF THE LEED REQUIREMENT

In an effort to reduce the amount of construction debris that eventually ends up in municipal landfills, the USGBC has established several credits for meeting their requirements for waste management. Essentially, the intent of the requirement is to encourage contractors to divert C&D debris from disposal in landfills and incinerators. This diversion can take the form of redirecting recovered resources back to the manufacturing process, that is, recycling. It can also be accomplished by redirecting reusable materials directly to other appropriate construction sites. The LEED requirement for management of construction waste on the jobsite is listed in the Materials and Resources section of the standard. It is called MR-2 construction and waste management.

LEED REQUIREMENTS FOR WASTE MANAGEMENT

To obtain one point in this credit, the contractor must undertake the following:

1. The contractor must divert, that is, recycle and salvage at least 50% of the nonhazardous C&D debris on the project. The amount of materials recycled can be calculated using the material's weight or volume. The only stipulation is that the method selected should be consistent throughout (USGBC). In addition, credit can be achieved if the percentage of diverted materials is increased to 75%.
2. The contractors must also develop and implement a construction waste management plan that, at a minimum, identifies the materials to be diverted from disposal. They must also determine if the materials will be sorted on-site or commingled. Hazardous materials, excavated soil, and land clearing debris do not contribute to this credit. Materials salvaged and reused on-site can be included in this calculation (USGBC).

TAKE NOTE

Remember to be successful:
- Prepare waste management plan
- Select materials for recycling
- Designate recycling storage areas on-site
- Identify local recyclers
- Document the process

STRATEGIES FOR SUCCESS

There are several ways the contractor can achieve the requirements for this credit; they include the following:

1. The contractor must prepare a waste management plan. This plan must include the establishment of goals for diversion from disposal in landfills and incinerators.

To achieve the associated LEED credit, the contractor must divert 50% of the construction waste traditionally sent to the landfill. This amount of material is calculated as a percentage of all the waste generated on-site during construction or demolition operations.

The waste management plan must be carefully prepared. It should anticipate the types of debris that will be generated by the project and establish specific goals for debris diversion. These goals must be able to be verified and documented. The following is a sample of a construction waste management plan.

2. The contractor should establish exactly which materials are going to be recycled. These might include cardboard, metal, brick, acoustical tile, concrete, plastic, clean wood, glass, gypsum wallboard, carpet, and insulation.

The contractor will initially select exactly which materials will be diverted from landfills or incinerators. The number of materials chosen is not as important as the potential volume these diverted materials represent. For instance, the selection of acoustical ceiling tile for diversion on a project that has little or no acoustic ceiling tile would not directly help in achieving the 50% quota. On the other hand, selecting gypsum wallboard and wood as

Waste Management Plan

Company: Northwest Best Construction
Project: Northwest Bank Building, Kent, WA
Designated Recycling Coordinator: John Doe

Waste Management Goals:
- This project will recycle or salvage for reuse xx% (e.g., 75%) by weight of the waste generated on-site.

Communication Plan:
- Waste prevention and recycling activities will be discussed at the beginning of each safety meeting.
- As each new subcontractor comes on-site, the recycling coordinator will present him/her with a copy of the waste management plan and provide a tour of the recycling areas.
- The subcontractor will be expected to make sure all their crews comply with the waste management plan.
- All recycling containers will be clearly labeled.
- Lists of acceptable/unacceptable materials will be posted throughout the site.

Expected Project Waste, Disposal, and Handling:
The following charts identify waste materials expected on this project, their disposal method, and handling procedures.

Example

Assume the contractor is filling a 30-cu. yd. waste container each month of an 18-month project. A 30–cu. yd. container will generally hold 5 tons of debris. So, each month 5 tons of construction debris is sent to the local landfill. Over the 18-month duration of the project, it would have generated 90 tons of C&D debris. To meet the LEED requirement, 50% or 45 tons of debris must be diverted from the landfill. This debris must be recycled or reused in some manner on the project.

materials to divert on a renovation project, which includes removal of all interior walls, would probably generate a high percentage of diverted debris. In addition, since the goal is a 50% diversion by weight, the contractor will be better served by analyzing the anticipated waste to be generated in relation to its weight. The following chart will be of assistance in undertaking an analysis to determine what materials might be best targeted for diversion (see Figure 6-12).

3. The contractor should designate a specific area(s) on the construction site for collection of recyclable materials. These areas can be segregated into separate containers or commingled into a single container. The decision on whether the separation of C&D waste will be undertaken on site or at the recycling depot will determine how many different C&D containers will be necessary.

The contractor, as part of his or her site logistics planning, must establish a convenient location to facilitate the collection and storage of construction debris which is intended to be diverted. The size of the dedicated collection area will vary with the types and amounts of diverted debris anticipated. In addition, the contractor, in conjunction with advice from local recycling haulers, must establish whether or not the debris will be separated or commingled. The process of diverting commingled waste varies from that of separated waste in that the separated waste is taken directly to the recycling facility for that material. Commingled waste, after leaving the site, is first taken to a sorting facility. At the sorting facility, the waste is separated into its components, which are then shipped to their respective recycling facilities. The contractor must establish which procedure is most effective in his or her individual locale.

4. The contractor must identify local construction haulers and recyclers capable of handling the types and volume of the designated materials.

The local "yellow pages" might be a first place to search for local construction debris recycling haulers. Another source for this information is The Bluebook of Building and Construction. This book contains a comprehensive regional listing of contractors, subcontractors and suppliers serving the construction industry. The Bluebook website is www.thebluebook.com. One of the best sources of construction waste recyclers is the National Institute of Building Sciences construction Waste Management Database. http://www.wbdg.org/tools/cwm.php. The database can be searched by state or zip code for recyclers of typical construction waste materials including the following:

- Appliances
- Asphalt
- Cardboard

Waste Management Diversion Summary			
Division / Recycling Materials Description	**Division / Recycling Hauler or Location**	**Quantity of Diverted / Recycled Waste**	**Units (tons or cy)**
Concrete	Rescoe Recycling	125	tons
Wood	Lumber	15	tons
Gypsum Wallboard	Rescoe Recycling	6.5	tons
Steel	Steel Reuse Industries	2.5	tons
Crushed Asphalt	On-Site Reuse	110	tons
Masonry	Rescoe Recycling	8	tons
Cardboard	Rescoe Recycling	1.5	tons
Total Construction Waste Diverted		268.5	tons
Landfill Materials Description	**Landfill Hauler**	**Quantity of Diverted / Recycled Waste**	**Units (tons / cy)**
General Mixed Waste	Lansing Landfill	55	tons
Total Construction Waste Sent to Landfill			55 tons
Total of All Construction Waste			323.5 tons
Percentage of Construction Waste Diverted From Landfill			83%

© CENGAGE LEARNING 2012

Figure 6-12

Waste management diversion summary.

- Carpet
- Ceiling Tile
- Concrete
- Gypsum Drywall
- Land Clearing/Soil
- Lighting
- Masonry
- Metals Ferrous
- Metals Non Ferrous
- Mixed/Commingled Waste
- Plastic
- Roofing: Asphalt-Based
- Roofing: EDPM
- Salvaged/Surplus Materials for Reuse
- Wood: Land Clearing Debris
- Wood: Scrap Lumber

5. The contractor must maintain good records of the recycling efforts throughout the construction process. Also, he or she must identify construction haulers and recyclers to handle the designated materials. These records of the materials diverted are essential in completing the required LEED template for the construction waste management credit. The following

Project: Johnson Limited Corporate Office Building		
Salvaged Reused Material Description	Source for Salvaged / Reused Material	Value / Product Cost ($)
Salvaged Brick	Bosworth Masonry	$65,000
Salvaged Steel Columns and Beams	Acme Steel	$45,000
Salvaged Wood Floor Joists	Imperial Wood Products	$35,000
Total Value of Salvaged Materials		$145,000
Total Construction Value (or 45% of default materials cost)		$1,565,000
Salvaged Materials as a Percentage of Construction Cost		

© CENGAGE LEARNING 2012

Figure 6-13

Salvaged materials chart.

Figure 6-14

Weight of construction waste per cubic yard.

Material	Lbs./Cu.Yd.
Wood	300
Drywall	500
Cardboard	100
Steel	1000
Rubble	0.5–1.3
Mixed Waste	350

COURTESY OF U.S. GREEN BUILDING COUNCIL–LEED REFERENCE GUIDE

chart illustrates how a contractor can track salvaged materials diverted from landfills (see Figure 6-13).

When calculating the weights of materials diverted, the contractor should use the following chart (see Figure 6-14).

6. Finally, the LEED protocol states that the diversion may include donation of materials to charitable organizations such as Habitat for Humanity and the reuse of salvaged materials on-site.

LEED DOCUMENTATION

After collecting and analyzing information on the waste diversion process the contractor is ready to upload the information to the LEED Online submittal form. The first step in this process is to establish the unit of measure for the C&D debris. The two choices for these units are tons or cubic yards. The second step of this process is to complete the interactive form on the website. If the contractor has carefully documented the C&D waste for the project as previously discussed in this chapter, uploading the information is a simple process. The resulting waste diverted from the landfills must be at a minimum 50% of the waste generated by the C&D operations, for one point and 75% for two points. The contractor must initial a statement certifying that to the best of their knowledge the information submitted is accurate.

REFERENCES AND SOURCES

There are many sources and references available for the constructor wishing to gain more information about diversion of materials from landfills and recycling. The internet provides a great resource. Listed below are several good sources for construction waste diversion information.

Recycling and Waste Management Councils

- Association of State and Territorial Solid Waste Management Officials, http://www.astswmo.org
- National Recycling Coalition, http://www.nrc-recycle.org
- Northeast Recycling Council, http://www.nerc.org
- Mid-America Council of Recycling Officials, http://www.marc.org
- Mid-Atlantic Consortium of Recycling and Economic Development Officials, http://macredo.org/
- Waste Reduction Resource Center, http://wrrc.p2pays.org

Recycling Associations

- Building Materials Reuse Association, http://www.bmra.org
- Construction Materials Recycling Association, http://www.cdrecycling.org
- National Demolition Association, http://www.demolitionassociation.com

Recycling Organizations

- The Deconstruction Institute, http://www.deconstructioninstitute.com
- GreenGoat, http://www.greengoat.org
- The Green Institute, http://www.greeninstitute.org
- Habitat for Humanity ReStore network, http://www.habitat.org/env/restores.aspx
- The Loading Dock, http://www.loadingdock.org
- Reuse Development Organization, http://www.redo.org
- Green Builder.com, www.greenbuilder.com/sourcebook/ConstructionWaste
- National Institute of Building Sciences, http://www.wbdg.org

USING MATERIALS THAT HAVE RECYCLED CONTENT

7

BACKGROUND

The construction of the built environment in the United States is an activity that consumes a considerable amount of the nation's resources. The manufacturing of these materials not only consumes valuable natural resources but the process can also add a significant amount of pollution to our atmosphere. One of the ways of reducing the impact on the environment, as discussed in Chapter 6, is to reduce the amount of construction waste going to landfills by diverting them to recycling facilities. Another way is to incorporate more materials with a high degree of recycled content into the construction process. This would serve to close the recycling circle. Unlike the previous chapter that discussed project waste management, this chapter explores the use of building materials that have recycled content. For instance, a contractor might use drywall, which has a high percentage of recycled content, or might use paint products with recycled content.

USE OF RECYCLED MATERIALS IN CONSTRUCTION

The use of recycled materials can have a considerable positive effect on preserving our nation's natural resources. According to the Earthworks Group, Americans discard 4 million tons of paper each year. The National Recycling Coalition estimates that recycling 1 ton of office paper saves 24 trees.[1] Not only does recycling save natural resources, but it also saves energy and reduces pollution. The energy required to manufacture a new product from a recycled one is often a fraction of that required to manufacture the product from a natural resource. For instance, the recycling of aluminum cans, the largest form of drink container, saves up to 95% of the energy required to manufacture a new can.[2]

This saved energy can be put to other uses. For instance, enough energy can be saved from the recycling of an aluminum can to supply an electric drill on the construction site for 1 h. In addition, it reduces 95% of air pollution generated by the manufacturing process. Glass is another commonly recycled material. Recycling 1 ton of glass saves approximately 1 ton of oil. The EPA estimates that making paper from paper goods that has been recycled will result in 74% less air pollution and 35% less water pollution than manufacturing paper from raw materials. Steel, one of the most commonly used construction materials, is also a prime material for recycling. Recycling steel reduces water and air pollution by 70%. It also consumes one quarter of the energy required to make new steel from virgin ore.

> **TAKE NOTE**
>
> Using recycled building materials results in:
> - Preserving natural resources
> - Reduced energy consumption
> - Reduced pollution

As stated in Chapter 1, the process of constructing and demolishing buildings is one of the largest contributors to the solid waste stream. Eventually much of this waste ends up in our nation's landfills. However, a large percentage of the C&D solid waste generated can be recycled. In recent years, many manufacturers that produce construction products and material have begun to incorporate recycled content in their products. These materials extend across the entire building process from structural materials like concrete and steel to finish materials like drywall and carpets. The following is a partial list of construction materials that could have recycled content:

- Concrete
- Concrete masonry units
- Structural steel
- Reinforcing steel
- Roofing
- Lumber
- Exterior sheathing
- Fiberboard
- Fiberglass insulation
- Flooring
- Wallboard
- Ductwork
- Ceramic tile
- Asphalt
- Paint
- Countertops

INTENT OF THE LEED REQUIREMENT

With the development of the LEED rating system for buildings, USGBC has been a major proponent of the use of materials with recycled content throughout the construction industry. The LEED credit for the use of materials with recycled content falls under the Materials and Resources section. It is listed as MR-4 Recycled

Content. MR-4 offers credits for buildings that incorporate materials with a significant degree of recycled content. The intent of this LEED standard is to increase demand for building products that incorporate recycled content materials, thereby reducing impacts resulting from extraction and processing of virgin materials.

LEED REQUIREMENTS FOR THE USE OF RECYCLED MATERIALS

As stated earlier, the LEED standards award credits for buildings that use materials with recycled content. The LEED requirement states that one credit will be awarded if 10% of the materials incorporated into a building are composed of recycled materials. An additional credit will be awarded if the recycled content for materials meets or exceeds 20% of the total materials incorporated into the project (see Figure 7-1).

Figure 7-1

LEED MR-4 credits.

Credits for Materials with Recycled Content	
Percentage of recycled content	Credits
10	1
20	2

© CENGAGE LEARNING 2012

Cost of Material Calculation

For the purposes of LEED credit calculations, the amount of materials incorporated into the project, which the LEED percentage is calculated upon, is defined as the cost of all materials in only the following CSI MasterFormat™ 2004 Edition Divisions:

Division 3 Concrete

Division 4 Masonry

Division 5 Metals

Division 6 Wood and Plastics

Division 7 Thermal and Moisture Protection

Division 8 Doors and Windows

Division 9 Finishes

Division 10 Specialties

Division 31.60.00 Foundations

Division 32.10.00 Paving

Division 32.30.00 Site Improvements

Division 32.90.00 Planting

The cost is defined by LEED as the actual cost of the materials including tax and cost of transportation, but exclusive of labor or installation cost.

$$\text{LEED material cost} = \text{material cost} + \text{tax} + \text{transportation cost}$$

This calculation can sometimes be difficult because some of the materials are supplied by subcontractors and those materials can be hard to track and document.

Subcontractors are often reluctant to open their books to the general contractor and allowing them to see what percentage of their costs are the materials and what percentage is labor.

Cost of Materials Alternative Calculations

If calculating the actual cost of materials proves to be too difficult, there is an alternative method of material cost calculation. The alternative method of determining the value of the materials in a project is to use a LEED-specified default value of 45% of the total project construction cost. The advantage of using this method of calculation is the reduced amount of documentation required. The contractor does not have to be concerned with a precise documentation of construction materials, the material cost will be assumed to be 45% of the total construction cost. However, the disadvantage is that projects with less than, in fact, have a material cost which is less than 45% of the total construction cost will have a more difficult time of achieving the 10% and 20% recycled content thresholds required to be awarded the LEED points. The contractor must then document the value of recycled materials and calculate this value as a percentage of the total value of materials as discussed earlier.

Example

For instance, in a reinforced masonry building, the building shell would probably consist of masonry, mortar, concrete grout, and steel reinforcing. The masonry contractor probably bid this installation as a lump sum to the general contractor. In order to properly document the material costs for the project, the masonry contractors would have to document their materials as a separate line item. In fact they should break out each individual component of material cost in the sub-bid. These numbers can then be used in calculating the total actual cost of materials incorporated into the project.

PostConsumer and PreConsumer Recycled Content

The LEED standard separates recycled content into two sub-categories, postconsumer recycled content and preconsumer recycled content. The standard accepts the definition of the recycled content of materials from the International Organization of Standards, ISO 14021, Environmental Labels and Declarations—Self-declared Environmental Claims.

- Postconsumer
  Postconsumer material is defined as a waste material generated by households or by commercial, industrial, and institutional facilities in their role as end-users of the product, which can no longer be used for its intended purpose. Recycled aluminum drink cans and used office paper are great examples of postconsumer recycled materials.

- Preconsumer

  Preconsumer material is defined as a material diverted from the waste stream during the manufacturing process. Excluded is reutilization of materials such as rework, regrind, or scrap generated in a process and capable of being reclaimed within the same process that generated it. This material never reaches the consumer. A good example of preconsumer content is steel scrap that might result from the manufacturing of cold-formed steel shapes. The cutting of the material

TAKE NOTE

Percentage of materials with recycled content can be based on either:
- Actual value of materials incorporated into the building or
- 45% of the total construction cost

Example

Assume that the flooring incorporated into the project has a recycled content and that this recycled content is divided into 10% postconsumer content and 85% preconsumer content. The value of the flooring is $45,000. By multiplying the value of the flooring ($45,000) times the preconsumer and postconsumer percentages, respectively, the pre and postconsumer values are determined. The sum of these two values equals the total value of the recycled content of the flooring (see Figure 7-2).

into its final shape often results in the creating of steel scraps that are left behind. These steel scraps are collected at the manufacturing facility and recycled prior to any consumer having contact with them. The weight of these scraps would be counted toward the percentage of preconsumer waste content included in the final product or material.

Within the standard, post- and preconsumer recycled content are not considered equal. The postconsumer material is counted at 100%, whereas the preconsumer content is taken at 50%. Therefore to meet this standard, the amount of postconsumer recycled content is added to one-half of the preconsumer content. The sum of the two must equal 10% (based on cost) of the total materials used in the project. An additional credit is achieved if the sum of both types of recycled materials equals 20% of the total value of the materials used in the project.

The following example illustrates a typical calculation to determine the recycled content value of steel in a construction project. It is similar to the calculation that would be required for each material with recycled content.

This calculation is repeated for each material that has recycled content in it. The sum of all of these values must be a minimum of 10% of the project's total material cost for one point and 20% for two points. The following table is an example of how the recycled content value is calculated on a project (see Figure 7-3).

Figure 7-2

Procedure to calculate recycled content value of steel.

Recycled content value (%)	= (% postconsumer recycled content × material cost) + 0.5 (% preconsumer recycled content × material cost)
Recycled content value (%)	= (10% × $45,000) + 0.5 (85% × $45,000)
Recycled content value (%)	= ($4,500) + 0.5 ($38,250)
$23,625	= ($4,500) + ($19,125)

© CENGAGE LEARNING 2012

Figure 7-3

Procedure to calculate total percentage of recycled content.

Material	Total material cost	Recycled content value	Total project material cost	% of recycled content
Structural steel	$45,000	$27,000	$575,000	4.5
Reinforcing steel	$115,000	$92,000	$575,000	16
Aggregate	$24,000	$24,000	$575,000	4.1
Gypsum Wallboard	$132,000	$119,000	$575,000	20
Particleboard	$3,000	$3,000	$575,000	0.5
Carpet	$43,000	$33,000	$575,000	5.7
Paint	$62,000	$12,500	$575,000	2.1
Total				52.9

© CENGAGE LEARNING 2012

Calculating Recycled Content of Building Assemblies

On occasion, the recycled material is a part of a building assembly. In this case, the first step in the recycled content calculation is to determine the percentages of the materials within the assembly. To determine the percentages required to meet the credit standard, the recycled content value of a material assembly must be determined. The percentage of recycled content material will be calculated by weight. First the weight of the recycled portion of the assembly is determined. This number is then multiplied by the cost of assembly to determine the recycled content value (see Figure 7-4).

Although most building products and assemblies can be used for this calculation, mechanical, electrical, and plumbing (MEP) components and specialty items such as elevators are specifically excluded from this calculation. In addition, only include materials permanently installed in the project can be counted. Furniture, if provided, can be included in this calculation, provided that it is included consistently in all the other calculations in the Material and Resource section of the LEED requirements.

Figure 7-4

Formula to calculate total project recycled content value.

$$\text{Assembly recycled content value (\%)} = \frac{\text{\% postconsumer recycled content}}{\text{Total assembly weight}} \times \text{Assembly cost} = 0.5 = \text{Assembly cost}$$

STRATEGIES FOR SUCCESS

During the design and construction documents phase, the contractors, in consultation with other members of the LEED compliance team, should establish a project goal for recycled content materials. They should then identify which materials in the project might most easily lead to the compliance goal. It is best to consider a range of environmental, economic, and performance attributes when selecting products and materials. The contractors must then identify material manufacturers and suppliers that can achieve this goal. During construction, the contractors must ensure that the specified recycled content materials are installed.

When considering recycled content the contractor should:

1. Step 1: Establish a project goal for recycled content materials.
   This goal will be determined by how many credits in this category the team is attempting to achieve. As stated earlier, one point will be achieved by incorporating building materials with recycled content value of 10%. Two points will be awarded for incorporating building materials with recycled content value of 20%.
2. Step 2: Select what materials are going to be targeted to have recycled content. While once very limited, the number of building materials that have recycled content is increasing annually. Look for the universal recycling symbol when looking for these materials. Sources for these materials are contained in Section 10.6 (see Figure 7-5).

Ultimately, the contractor must document that the building materials do in fact have recycled content and what the percentage of recycled content is. To assist the contractor, there are independent agencies that certify the recycled content of building materials. One such agency is the Scientific Certification Systems (SCS). The SCS was launched in 1993 and has become one of the foremost sources of building materials recycled content certification. The certification addresses both preconsumer and postconsumer content as is required by the LEED standard. The recycled content of materials certified by the SCS is accepted by USGBC (see Figure 7-6).

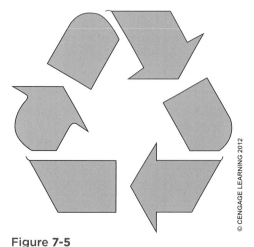

© CENGAGE LEARNING 2012

Figure 7-5

Recycling symbol.

COURTESY OF SCIENTIFIC CERTIFICATION SYSTEMS

Figure 7-6

Scientific Certification Systems (SCS).

LEED DOCUMENTATION

In the first step in the LEED Online documentation of compliance with this credit, the contractor must select whether the basis of cost will be the actual cost of all materials or the cost will be based on the default 45% of the total construction cost. The contractor will check off the appropriate box in the form. The next step involves the contractor selecting which credits are being attempted. These include the following:

- MRc3: Materials Reuse.
- MRc4: Recycled Content.
- MRc5: Regional Materials.
- MRc6: Rapidly Renewable Materials.
- MRc7: Certified Wood.

The next step of the online documentation process involves uploading to the LEED Online Web site all of the information required to document compliance with the requirements for this credit. This includes completing the online submittal recycled materials table. Again if the contractor has carefully documented the percentages of recycled materials as previously discussed in this chapter the completion of this step is fairly easy.

The last page of the online form requires the contractor to check off a certification that all of the recycled materials claimed in the Recycled Content table meet the ISO 14021 definitions of post and preconsumer material. In addition, the contractor must certify that MEP materials were excluded from materials in the Recycled Content table.

This credit like several others offers an expedited documentation process. In this case, the option is available of a licensed architect or interior designer is making the submittal. If the contractor is making the submittal, additional documentation is required. The documentation required is in the form of material and product cut sheet uploaded for a minimum of 20% of the materials with recycled content.

REFERENCES AND SOURCES

International Organization of Standards document ISO 14021-1999: Environmental Labels and Declarations—Self-declared Environmental Claims

BuildingGreen, Inc. Greenspec at http://www.buildinggreen.com. This site provides listing of over 2000 green building products. Each listing has full information on the product including manufacturer information and environmental data.

California Integrated Waste Management Board, Recycled Content Product Directory At http://www.ciwmb.ca. This list also provides a listing of environmentally friendly buildings materials that have recycled content.

Center for Resourceful Building Technology, Guide to Resource-Efficient Building Elements at http://www.cbrt.ncat.org. This is another site that provides resources on building materials with recycled content.

Oikos at http://oikos.com. This site serves as a good resource for sustainable building products including materials with recycled content.

U.S. EPA Comprehensive Procurement Guidelines Program at http://www.epa.gov/cpg/products.htm. This EPA site has the EPA procurement guideline that has a wealth of recycled materials information.

Construction Specifications Institute (CSI), Green Format at http://www.greenformat.com. This site lists the CSI green format specifications as well as listing sustainable building products.

8

REGIONAL MATERIALS

BACKGROUND

All building materials have what is referred to as embodied energy. The term embodied energy refers to the energy consumed in the manufacture, transportation, and installation of materials into a building. There are two primary classifications of embodied energy, initial embodied energy and recurring embodied energy.

1. Initial Embodied Energy

   Initial embodied energy includes the energy consumed in the acquisition of the raw materials, the manufacturing process itself, the energy used to transport the finished material from the factory to the site, and the energy required to incorporate the material into the building. This initial embodied energy has two components, direct energy and indirect energy. Direct energy is the energy used in the transportation process from the point of manufacturer to the project site. The indirect energy is the energy used in the raw materials acquisition and manufacturing process. This would include the energy used in the transportation phase of that process.

2. Recurring Embodied Energy

   This term refers to the energy consumed in the maintenance, repair, or replacement of materials and components already installed within a building. This would be tracked for the life of the building.

TAKE NOTE

Energy in building materials

- Initial embodied energy
- Recurring embodied energy

MEASURING EMBODIED ENERGY

The embodied energy of a material is generally measured as a quantity of nonrenewable energy required in the manufacturing, delivery, and installation process per unit of building material. It is usually expressed as megaJoules (MJ) or gigaJoules (GJ) per unit of weight or area. A megaJoule is equal to 1 million Joules or is approximately the kinetic energy of a 1 ton vehicle moving at approximately 100 mph. A gigajoule equals 1 billion Joules or is roughly equal to the energy in one-sixth of a barrel of oil. The calculation of this embodied energy is a complex process that involves many sources of data. In addition to being a pure measurement of energy consumption, it does include the environmental implementations of greenhouse gas production and the depletion of natural resources.

The amount of embodied energy a specific material contains is in fact one measure of its impact on the environment. For instance, a material like aggregate used for concrete has an embodied energy of 0.10 MJ/kg, whereas a PVC pipe would has an embodied energy of 70 MJ/kg. This means that the manufacturing of the PVC pipe has an impact on the environment of 700 times than that of the aggregate for the same quantity of material. Plywood has an embodied energy of 10.4, which is approximately 10.5 times than that of concrete block, which has an embodied energy of 0.94 MJ/kg.

The following chart illustrates the embodied energy in commonly used building materials. The lower the number, the less impact the manufacturing of the material has on the environment (see Figure 8-1):

When we consider the total building structure, we find that 50% of the embodied energy of the building is in the construction of the structure and the building envelope. The following chart illustrates the embodied energy of a typical office building (see Figure 8-2).

Therefore the primary focus of the embodied energy of the materials should be on the building's structure and envelope. The embodied energy in just these two components of the building constitutes over 50% of the total embodied energy in the building.

Economic Considerations

The environmental impact of transporting construction materials long distances has been fairly well-documented. In addition to those environmental factors, there are also economic considerations. The cost of transporting materials from the point of the harvesting of the raw materials to the point of product manufacturing and then the transportation of that finished product to the jobsite can often be a considerable expense.

The Recycled Materials Resource Center, RMRC, states that the cost of any construction material or product is roughly composed of five different components. They are:

1. The price of the raw material (Pr)
   This cost is the actual cost of the unprocessed raw material. The cost is established by the raw material generator and might vary according to the dynamics of supply and demand.

Figure 8-1

Embodied energy in construction materials.

MATERIAL	EMBODIED ENERGY	
	MJ/kg	MJ/m³
Aggregate	0.10	150
Straw bale	0.24	31
Soil-cement	0.42	819
Stone (local)	0.79	2030
Concrete block	0.94	2350
Concrete (30 Mpa)	1.3	3180
Concrete precast	2.0	2780
Lumber	2.5	1380
Brick	2.5	5170
Cellulose insulation	3.3	112
Gypsum wallboard	6.1	5890
Particle board	8.0	4400
Aluminum (recycled)	8.1	21870
Steel (recycled)	8.9	37210
Shingles (asphalt)	9.0	4930
Plywood	10.4	5720
Mineral wool insulation	14.6	139
Glass	15.9	37550
Fiberglass insulation	30.3	970
Steel	32.0	251200
Zinc	51.0	371280
Brass	62.0	519560
PVC	70.0	93620
Copper	70.6	631164
Paint	93.3	117500
Linoleum	116	150930
Polystyrene insulation	117	3770
Carpet (synthetic)	148	84900
Aluminum	227	515700

NOTE: Embodied energy values based on several international sources-local values may vary.

2. The cost of processing the material (Cp)
   Almost all construction materials require some degree of process to make the virgin raw materials suitable for use in the final material or product. Included in this cost would be an incremental cost of equipment and labor used ion the processing operation.
3. The cost of stockpiling the material (Cs)
   Most if not all construction materials are stockpiled for a period of time prior to its final use. The duration of this time will be dependant on the

Figure 8-2

Breakdown of embodied energy in buildings.

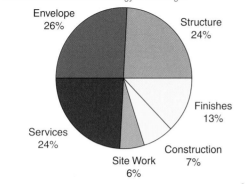

Envelope 26%
Structure 24%
Finishes 13%
Services 24%
Construction 7%
Site Work 6%

COURTESY OF CANADIAN ARCHITECT

Average Total Initial Embodied Energy 4.82 GJ/m²

demand of the given materials as compared with the given amount stockpiled. The cost associated with stockpiling will be directly related to the amount of time the material is stockpiled.

4. The cost of loading the material (Cl)

When the material is ready to be delivered to the jobsite, it must first be loaded onto the transportation vehicle. This could be a truck, a train, or a combination of both. Again there is commonly an incremental cost for loading equipment and labor included in this cost.

5. Cost of transporting the material (Ct)

The next component of construction material cost is the cost of the transportation of the material from the point of manufacturer or point of stockpiling to the jobsite. This cost will vary depending on the mode of transportation selected, that is, truck, train, air, and the distance the material is to be transported.

The final cost of the material from its generation as a raw material to the delivery in its final form on the jobsite can be illustrated by the following equation:

$$Pr + Cp + Cs + Cl + Ct = \text{final material cost}$$

$$\text{raw material cost} + \text{processing cost} + \text{stockpiling cost} + \text{loading cost} + \text{transportation cost}$$

INTENT OF THE LEED REQUIREMENT

The intent of the LEED MR 5 credit is to reduce the impact of construction on the environment. In particular, the credit serves to encourage the use of locally obtained materials, because the embodied energy in materials obtained locally is much lower, when compared with materials that have to be shipped long distances. Hopefully this requirement will lead to an increased demand for locally produced materials and products.

LEED REQUIREMENTS FOR REGIONAL MATERIALS

Credit MR 5 Regional Materials addresses the use of regionally obtained materials in the building. The credit requires that the contractor uses building materials or products that have been extracted, harvested, or recovered, as well as manufactured, within 500 miles of the project. For LEED purposes, the point of manufacturing is considered the final point of the product's or assembly's manufacturing or assembly prior to delivery to the site.

Figure 8-3

LEED MR5 credits.

Credits for Regional Materials	
Percentage of Regional Materials	Credits
10%	1
20%	2

© CENGAGE LEARNING 2012

According to USGBC LEED version 3.0 standard, "mechanical, electrical, and plumbing components and specialty items such as elevators and equipment must not be included in this calculation. Include only materials permanently installed in the project can be considered. Furniture may be included if it is included consistently in MR Credit 3: Materials Reuse through MR Credit 7: Certified Wood."

STRATEGIES FOR SUCCESS

Step 1

The first step in fulfilling the requirements for this credit is for the design/ construction team to establish a project goal for locally sources materials and products. This will be either 10% of the project material value to achieve one point or 20% to achieve two points. The decision to attempt one or two points will probably be made after a preliminary analysis of locally derived materials is undertaken. There is no point in attempting to achieve two points that would require 20% of the materials to be obtained locally if that quantity of materials is not readily available locally. The failure to achieve anticipated points will require additional points to be achieved in other categories in order to meet the anticipated level of LEED award.

Example

For instance, if the component being considered is the roof trusses, roof trusses are generally composed of two materials. They are composed of wood members and metal plates. Say the wood members come from lumber mills in Macon, Georgia, the metal plates are manufactured in Southern California and the trusses are assembled in Jacksonville, Florida. Jacksonville would be considered the point of manufacture. To be considered as regionally obtained materials the project site must be within 500 miles of the Jacksonville, Florida truss plant.

To meet the LEED credit requirement, the value of these all the regionally obtained materials must be equal to or exceed at least 10% of the value of the projects materials for one point or 20% to be awarded two points. The minimum percentage of regional materials incorporated into the project for each point threshold is as follows (see Figure 8-3):

Step 2

The second step will be to determine the cost basis of the materials to be incorporated into the project. This will generally be the total cost of all materials in the following CSI divisions (see Figure 8-4):

Figure 8-4

CSI divisions contributing to MR5 credits.

Contributory Costs for Regional Materials Credit	
CSI Division	**Description**
03	Concrete
04	Masonry
05	Metals
06	Wood and Plastics
07	Thermal and Moisture Protection
08	Openings
09	Finishes
10	Specialties
31.60	Earthwork: Foundations
32.10	Exterior Improvements: Paving
32.30	Exterior Improvements: Site Improvements
32.90	Exterior Improvements: Planting

© CENGAGE LEARNING 2012

The following chart illustrates the divisional breakdown on a branch bank facility located in Miami, Florida (see Figure 8-5):

ABC Construction					
	Division	**Material Cost**	**Labor Cost**	**Equipment Cost**	**Subtotal Cost**
Division 1	General Conditions	$ 65,223.38	$ 27,445.75	$ 546.52	$ 93,215.65
Division 2	Site Work	$ 62,567.82	$ 54,986.54	$ 16,894.65	$ 134,449.01
Division 3	Concrete	$ 92,546.34	$ 138,587.64	$ 15,487.65	$ 246,621.63
Division 4	Masonry	$ 25,876.98	$ 52,365.25	$ 15,695.32	$ 93,937.55
Division 5	Metals	$ 27,856.54	$ 14,236.89	$ 14,562.24	$ 56,655.67
Division 6	Woods and Plastics	$ 9423.76	$ 7551.46	$ -	$ 16,975.22
Division 7	Thermal and Moisture Protection	$ 16,756.48	$ 13,568.41	$ 1577.74	$ 31,902.63
Division 8	Doors and Windows	$ 65,889.54	$ 9,785.25	$ 154.00	$ 75,828.79
Division 9	Finishes	$ 54,829.43	$ 55,689.54	$ 475.32	$ 110,994.29
Division 10	Specialties	$ 6297.56	$ 550.30	$ 2658.98	$ 9506.84
Division 14	Conveying Systems	$ 47,223.74	$ -	$ -	$ 47,223.74
Division 15	Mechanical	$ 59,778.43	$ 35,465.25	$ 1800.00	$ 97,043.68
Division 16	Electrical	$ 79,657.47	$ 72,659.84	$ -	$ 152,317.31
	Total	$ 613,927.47	$ 482,892.12	$ 69,852.42	$1,166,672.01

© CENGAGE LEARNING 2012

Figure 8-5

Sample divisional breakdown.

Step 3

Using the total cost established in Step 2, the contractor should determine the target value in dollars of the percentage target selected to be achieved by the design/construction team. This can be calculated using the following formula (see Figure 8-6):

Figure 8-6

Formula to calculate the target value of local materials in the project.

Target Percentage of Local Materials	×	Total Cost of all Materials in Divisions 03-10 + 31.60 + 32.10 + 32.30 + 32.90	=	Target Value of Locally Sourced Materials

© CENGAGE LEARNING 2012

In the case of our example the sum of the materials in Divisions 3–10 is as follows (see Figure 8-7):

Figure 8-7

Sample materials cost breakdown.

	Division	Material Cost
Division 1	General Conditions	$ 65,223.38
Division 2	Site Work	$ 62,567.82
Division 3	Concrete	$ 92,546.34
Division 4	Masonry	$ 25,876.98
Division 5	Metals	$ 27,856.54
Division 6	Woods and Plastics	$ 9423.76
Division 7	Thermal and Moisture Protection	$ 16,756.48
Division 8	Doors and Windows	$ 65,889.54
Division 9	Finishes	$ 54,829.43
Division 10	Specialties	$ 6297.56
	Total	$ 427,267.83

© CENGAGE LEARNING 2012

In order to achieve one LEED credit, 10% of the materials must be obtained regionally. To obtain this credit, the contractor must obtain a total material value of $42,727.00, which is 10% of the total calculated material cost listed above, from a local source. To be awarded two credits, the total value of locally obtained materials must equal 20% or $85,454.00.

Step 4

Once the regional material value targets are identified, the contractor should develop a detailed plan of how this target value will be achieved. To do this the contractor must undertake a division by division analysis. This analysis will involve identifying which materials in each division can be readily obtained within the 500 mile parameter established in the LEED requirement.

Sometimes a regionally obtained material is a part of a building component that may not comply in its entirety with the 500 mile requirement. If portions of materials or products have locally obtained materials, these can be calculated based on the weight of the locally obtained materials as a percentage of that materials total weight. The following formulas can be used to calculate partial value of locally obtained materials (see Figure 8-8):

Figure 8-8

Formula to calculate percentage of partial local materials component.

Weight of Local Materials in the Final Product / Weight of The Final Product	$\times$	100	$=$	Percentage of Locally Sourced Material in that Product

© CENGAGE LEARNING 2012

For example if poured in place concrete contains regionally obtained rock as aggregate then the weight of the aggregate as a percentage of the total concrete weight can be multiplied by the total cost of the concrete to obtain the dollar value of the regional material component. The following chart illustrates a calculation of regionally obtained material for concrete (see Figure 8-9):

Figure 8-9

Calculation of regionally obtained materials in composite assemblies.

Component	Weight (lbs)	Distance between Projects and extraction/ Manufacturing point	Weight (lbs) contributing to Regional Materials credit
Cement	423	1500	0
Fly Ash	423	1575	0
Water	413	7	413
Slag	1125	1550	0
Course Aggregate Rock	1500	35	1500
Fine Aggregate Sand	1800	12	1800
Component Totals	5684		3,713
Percentage of Regionally Obtained Material in Concrete			65.3%

© CENGAGE LEARNING 2012

In this example 65.3% of the concrete, by weight is obtained locally within the 500 mile parameters of the LEED credit. This percentage can then be multiplied by the total project cost of concrete to determine the dollar value of the regionally obtained concrete. The following formula can be used to undertake this calculation (see Figure 8-10):

Figure 8-10

Formula to calculate target value of partial local materials components.

Percentage of Local Materials in the Final Product (based on weight)	$\times$	Total Cost of that Material	$=$	Value of Locally Sourced Material

© CENGAGE LEARNING 2012

As previously determined in our example, 65.3% of the concrete is obtained locally. When multiplied by the total cost of the concrete material or $92,546.34 the dollar value of the regionally obtained concrete ($60,432.76) is determined. This value for regionally obtained concrete when compared with the total material cost of $418,267.83 equals 14.45%. The use of regionally obtained materials in Division 3 alone qualifies the project for one LEED credit. In order to achieve the second credit, the contractor must obtain an additional 5.55% or $5090.00 in locally obtained materials.

Step 5

In the fifth and final step of this process, the contractor must purchase the materials and provide documentation to verify that the materials do indeed comply with the LEED requirement.

LEED DOCUMENTATION

In the first step in the LEED Online documentation of compliance with this credit for regional materials, the contractor must select whether the basis of cost will be the actual cost of all materials or the cost will be based on the default 45% of the total construction cost. The contractor will check off the appropriate box on the form. The next step involves the contractor selecting which credits are being attempted. These include the following:

- MRc3: Materials Reuse
- MRc4: Recycled Content
- MRc5: Regional Materials
- MRc6: Rapidly Renewable Materials
- MRc7: Certified Wood.

The contractor is also required to check off whether or not furniture is included in the calculation. The second step in this process is for the contractor to upload documentation on compliance with the regional materials credit to the form on the LEED Online Web site. This is followed by the contractor checking off that MEP materials were excluded from materials in the Regional Materials table. In addition, materials that were used to contribute to MRc1 were excluded from the Regional Materials table.

This credit like several others offers an expedited documentation process. In this case, the option is available of a licensed architect or interior designer is making the submittal. If the contractor is making the submittal, additional documentation is required. The documentation required is in the form of material and product cut sheet uploaded for a minimum of 20% of the regionally acquired materials.

RAPIDLY RENEWABLE MATERIALS

BACKGROUND

As discussed in Chapter 1, the process of constructing our nation's homes, offices, and schools has a significant effect on our nation's natural resources. In fact, for several decades the construction industry has been one of the largest users of natural resources in the United States. The depletion of our nation's natural resources through construction activities can be significantly reduced through proper design and construction planning of nation's buildings.

TAKE NOTE

There are two classifications of natural resources
- Nonrenewable resources
- Renewable resources

NATURAL RESOURCES

NonRenewable Resources

To look at the issue of the usage of natural resource, we must first distinguish between the two major classes of natural resources. The first classification is non-renewable natural resources. These are naturally occurring resources that while seemingly abundant are actually in limited supply and not renewable. Once the finite supply is used there will be no more material available except that which might be recycled. Examples of these nonrenewable resources would include the following:

- Stone
- Sand
- Petroleum products

Renewable Resources

The second classification of natural resources is the renewable resources. Although there are technically many renewable resources, LEED is focusing on rapidly renewable resources. These resources have a quick turn around and can be harvested soon after planting. For instance wood, in order to be considered for LEED credit, must come from trees that have a harvesting cycle of 10 y. or less. If animal-based, the material must be able to be harvested without harming the animal. Wool that can be sheared off without harming the animal is a renewable resource. However, leather is not. Examples of renewable natural resources include the following (see Figure 9–1):

Figure 9–1

Rapidly renewable resource harvesting cycle.

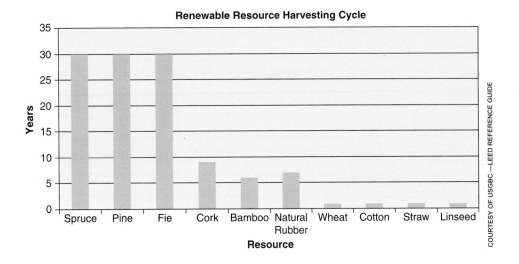

- *Wood*
  Although all wood is considered a natural resource, not all wood can be considered to be rapidly renewable. As mentioned earlier, in order to be considered rapidly renewable, the resource must be able to be harvested on a maximum 10-y. interval. Some of the following types of wood will comply with this requirement.
- *Bamboo*
  Bamboo is rapidly becoming one of the fastest growing rapidly renewable resources in the United States' construction industry. It can be harvested within 5 y. of planting without affecting the overall canopy. Bamboo started out as a finish material in laminated flooring but it is now used in a variety of products ranging from furniture to wall coverings.
- *Natural rubber*
  Natural rubber is manufactured from the sap of the rubber tree. This tree originated in Brazil but the imported materials now primarily come from Thailand, Indonesia, Malaysia, and Sri Lanka. The sap is harvested without harming the trees. The harvested sap is then used to make a multitude of products. However, in the construction industry the use of natural rubber is generally limited to flooring products.

- *Cork*

  It was once thought that this natural resource was dwindling toward extinction. However, this is not true. Cork is a product derived from the cork oak tree. Unlike most trees that are severely harmed when their bark is removed, this tree does not suffer any harm. The tree must first be approximately 25 y. old. After that age, the bark can be stripped every 8–9 y. Stripped bark is formed into sheets. These cork sheets are primarily used in flooring and wall covering products.

- *Cotton*

  Cotton, of course, is a naturally grown product that is easily regenerated. Cotton is currently used in a variety of building products mostly in the area of building insulation.

- *Wool*

  Wool can be easily shone from sheep about every 2 y. without harming the animal. It is a rapidly renewable material that has been used in the making of fabrics for over a thousand years. In the building industry, wool is primarily used in the manufacturing of carpets.

- *Linoleum*

  Linoleum is a linseed oil-based product employed primarily in the flooring industry. Although linoleum was once a very popular flooring material in the 1930s, 1940s, and 1950s, it gave way to the expanding use of vinyl flooring and base. Now in an effort to rid ourselves of these chemically laden products, natural materials such as linoleum are again being used.

- *Agrifiber boards*

  These rapidly renewable materials also discussed in Chapter 16 include strawboard and wheatboard. These materials are generally pressed into boards for use in cabinetry and furniture construction.

The following chart discusses how these products are used in the construction industry and compares their advantages and disadvantages (see Figure 9–2):

TAKE NOTE

Common renewable resources
- Bamboo
- Cork
- Natural rubber
- Cotton
- Wool
- Linoleum
- Agrifiber products

PRODUCT	TYPICAL BUILDING APPLICATIONS	ADVANTAGES	DISADVANTAGES
Bamboo	• Finish flooring • Cabinetry • Veneers	Wears well, readily available	Can shrink and bend if it is not properly acclimated to the building's indoor environment.
Natural rubber	• Flooring	Durable and easy to install. Can be formed or cut into a variety of shapes.	Considerably more expensive than the synthetic materials such as vinyl.
Cork	• Finish flooring • Wall coverings • Carpet padding	Easy to install, resilient and resistant to denting. It can be cut into a variety of shapes.	It is not moisture resistant and will swell if subjected to moisture or high humidity. It is not colorfast and will fade if placed in direct sunlight.
Cotton	• Thermal insulation • Sound Insulation	Easy to install has good thermal and sound resistant characteristics. Can be easily recycled.	Is generally more expensive than comparable batts of fiberglass.
Wool	• Carpeting	Soft and relatively durable.	Often considerably more expensive that other similar fibers.
Linoleum	• Sheet flooring and base	Easy to install, very durable with high wear resistance, comes in a variety of colors and can be cut into multitude of shapes.	It is considerably more expensive than its counterpart, vinyl flooring.
Agrifiberboards	• Cabinetry • Furniture • Door cores	Can be used as a sheet product or be used as individual pieces similar to nominal lumber.	Can be environmentally harmful if urea formaldehyde is used as a binder.

© CENGAGE LEARNING 2012

Figure 9–2

Advantages and disadvantages of rapidly renewable materials in construction.

INTENT OF THE LEED REQUIREMENT

The intent of the LEED MR6 credit requirement is to promote a reduction in the depletion of the world's natural resources due to construction operations. To do this the LEED standard rewards a project credit for using rapidly renewable resources in place of resources that are either in finite supply or are slow to be naturally regenerated. The credit for the use of rapidly renewable resources is contained in Credit 6 of the Material Resources section of the LEED standards, MRc6.

LEED REQUIREMENTS FOR RAPIDLY RENEWABLE MATERIALS

The standard for the use of rapidly renewable materials on buildings projects is defined in LEED MR Credit 6. The standard requires that in order to receive one credit point, projects must incorporate "rapidly renewable materials" into the project. These rapidly renewable materials are defined by the standard as naturally grown materials that can be harvested on a maximum 10-y. cycle. This means that the materials reach growth maturity and can be fully harvested within 10 y. of planting. The value of these materials must be at least 2.5% of the total of all building materials and products used in the building. This percentage is based on the cost of the rapidly renewable materials as compared with the total cost of all materials incorporated into the building (see Figure 9–3).

Figure 9–3

LEED MR-6 credits.

Credits for Using Rapidly Renewable Materials	
Percentage of Rapidly Renewable Materials	Credits
2.5%	1
5%	2

© CENGAGE LEARNING 2012

STRATEGIES FOR SUCCESS

In order to successfully obtain this credit, the project team must first establish a project goal for rapidly renewable materials. The contractors must identify which products they will use to comply with the requirements for this credit. Once the products have been chosen, suppliers who can support the achievement of this goal must be identified. As compliance only requires 2.5% of the total value of construction materials incorporated into the project, the actual compliance is not very hard. The key to compliance is to start identifying the rapidly renewable resources prior to commencement of construction and track compliance from the beginning. Once the actual materials and products are selected the contractor must make sure that these products are in fact compliant and are actually installed in the quantities anticipated. To achieve this, the following form can be used to track and document compliance with this credit (see Figure 9–4).

As discussed earlier, the value of materials incorporated into the project can be either the actual cost of all materials excluding labor and equipment or 45% of the total cost of construction.

The contractor must obtain documentation from the product manufacturer that the product does in fact consist of rapidly renewable materials. If the contractor fails to obtain this certification prior to the installation and later finds out that the product does not contain rapidly renewable materials, the points for this credit might be lost. On the following page is a sample certification from an insulation manufacturer that used rapidly renewable materials.

Figure 9–4

Sample renewable resource chart.

Summit Office Building Renewable Resources				
Total Construction Cost (or)				
Actual Cost of All Materials				**$420,000**
Product Name	**Vendor Name**	**Assembly Product Cost**	**% of Rapidly Renewable Materials**	**Value of Rapidly Renewable Materials**
Wheatboard Cabinetry	Johnson Brothers Cabinetry	$15,000	30%	$4500
Linoleum Flooring	Acme Carpet and Flooring	$12,000	50%	$6000
Bamboo Flooring	Acme Carpet and Flooring	$32,000	90%	$28,800
Bamboo Window Blinds	Sonnet Window Coverings	$7500	75%	$5625
Total Value of Recycled Materials				$44,925
Percentage of Recycled Materials				10%
Points Documented				2

From: Sean Desmond
            24053 S Arizona Ave Ste 151
            Chandler, AZ 85248
            480.812.9114 – Phone

Subject: UltraTouch™ Denim Insulation LEED Cert

To Whom It May Concern:

Please find attached the requested information regarding the recycled content for Bonded Logic's insulation products as well as our rapidly renewable material resources usage. All percentages are calculated by weight.

RECYCLED CONTENT

Product	% Post Consumer	% Pre Consumer
UltraTouch™ Denim Insulation	90%	0%

RAPIDLY RENEWABLE RESOURCES

Product	Raw Material Source	% Utilized
UltraTouch™ Denim Insulation	Cotton	80%

Material Sourcing/Manufacturing Info

Manufacturing Location – Chandler, AZ 85248

Extraction Site – Brownsville, TX 78520

Signature:
Job Title: Sales/Marketing Manager
Company Name: Bonded Logic, Inc.

CERTIFIED WOOD

BACKGROUND

As mentioned in Chapter 1, the activities undertaken by the construction industry make a significant impact on our environment. One of the major impacts created by the industry is on our nation's forests. Before the advent of the settlers arriving on this land primarily from Europe, there was approximately 1 billion acres of forest in what is now called the United States.[1] Currently there is approximately 747 million acres in the contiguous United States, not including Alaska or Hawaii. This forest area is divided almost equally between the eastern forest, which covers approximately 384 million acres and the western forest, which covers approximately 363 million acres.

As this forest area is lost, not only do we loose a valuable natural resource that serves as a source of wood for our homes and commercial buildings but also reduce the natural habitat for many animal species. As the natural habitat is lost, many animal species in the United States are displaced and are forced to repeatedly move. This has resulted in several species of animals that were once in abundant numbers being classified as endangered. Animals are generally well-adapted and able to change as their environment changes gradually over time. However, most species are unable to adequately adapt to rapid changes to their habitat. It is this rapid change to our nation's natural habitat that is the greatest source of endangerment to animals within the United States.[2]

In addition, this deforestation results in a considerable increase in the potential for erosion. This erosion eventually results in a significant increase in the pollution of our nation's rivers, lakes, and streams.

Figure 10-1

FSC-certified wood stamp.

THE FOREST STEWARDSHIP COUNCIL

Established in 1993, the Forest Stewardship Council (FSC) was created to effect a change on the process of the practice of forestry worldwide. As stated by the FSC (see Figure 10-1),

FSC sets forth principles, criteria, and standards that span economic, social, and environmental concerns. The FSC standards represent the world's strongest system for guiding forest management toward sustainable outcomes. Like the forestry profession itself, the FSC system includes stakeholders with a diverse array of perspectives on what represents a well-managed and sustainable forest.[3]

The FSC sets forth certain standards for forestry operations, which will reduce the impact on the environment. It is guided by 10 principles that range from the rights of indigenous people living in and around the forests to environmental policies and forestry management. The following is a listing of the 10 guiding principles of the FSC.

Principle 1: Compliance with Laws and FSC Principles

Forest management shall respect all applicable laws of the nation in which they occur, and international treaties and agreements to which the nation is a signatory, and comply with all FSC Principles and Criteria.

Principle 2: Tenure and Use Rights and Responsibilities

Long-term tenure and use rights to the land and forest resources shall be clearly defined, documented, and legally established.

Principle 3: Indigenous Peoples Rights

The legal and customary rights of indigenous peoples to own, use, and manage their lands, territories, and resources shall be recognized and respected.

Principle 4: Community Relations and Workers Rights

Forest management operations shall maintain or enhance the long-term social and economic well-being of forest workers and local communities.

Principle 5: Benefits From the Forest

Forest management operations shall encourage the efficient use of the forest's multiple products and services to ensure economic viability and a wide range of environmental and social benefits.

Principle 6: Environmental Impact

Forest management shall conserve biological diversity and its associated values, water resources, soils, and unique and fragile ecosystems and landscapes, and, by so doing, maintain the ecological functions and the integrity of the forest.

Principle 7: Management Plan

A management plan—appropriate to the scale and intensity of the operations— shall be written, implemented, and kept up to date. The long-term objectives of management, and the means of achieving them, shall be clearly stated.

Principle 8: Monitoring and Assessment

Monitoring shall be conducted—appropriate to the scale and intensity of the forest management—to assess the condition of the forest, yields of forest products, chain of custody (COC), management activities, and their social and environmental impacts.

Principle 9: Maintenance of High Conservation Value Forests

Management activities in High Conservation Value Forests shall maintain or enhance the attributes that define such forests. Decisions regarding High Conservation Value Forests shall always be considered in the context of a precautionary approach.

Principle 10: Plantations

Plantations shall be planned and managed in accordance with Principles and Criteria 1–9, and Principle 10 and its Criteria. Although plantations can provide an array of social and economic benefits, and can contribute to satisfying the world's needs for forest products, they should complement the management of, reduce pressures on, and promote the restoration and conservation of natural forests.

FSC CERTIFICATION

From a sustainable building standpoint, the FSC serves to accredit third-party certification organizations. These third-party agents are tasked to annually evaluate companies for compliance with the FSC standards. The following is a listing of FSC third-party certification agencies.

FSC-ACCREDITED CERTIFIERS WITH OFFICES IN THE UNITED STATES

1. American Green, a partner of GFA consulting group
   Web site: www.americangreen.net
   American Green provides personalized FSC COC service nationwide, including Hawaii and Alaska

2. Bureau Veritas Certification
   Web site: www.us.bureauveritas.com/bvc

3. QMI-SAI Global
   Web site: www.qmi-saiglobal.com

4. SGS Systems & Services Certification USA
   Web site: www.us.sgs.com/forestry_us
   SGS provides service to all 50 states and has auditors throughout the nation. For useful information, and inquiries, please contact a full-time representative for the USA.

5. Rainforest Alliance/Smartwood Program
   Web site: www.rainforest-alliance.org/smartwood

6. Scientific Certification Systems, Inc.
   Web site: www.scscertified.com SCS centrally administers a worldwide program for both forest management and COC. We have clients in all 50 states and field auditors with offices in the following states: Alaska, Arizona, California, Colorado, Connecticut, Georgia, Maine, Maryland, Massachusetts, Michigan, Minnesota, Mississippi, Montana, New Hampshire, Ohio, Oregon, Pennsylvania, Tennessee, Texas, Vermont, Virginia, Washington.

7. Soil Association—Woodmark
   Web site: www.soilassociation.org/forestry Forest Management and Chain-of-Custody certification

Within the FSC there are two types of certifications. The first certification addresses compliance within the forest, whereas the second certification addresses the process from forest to installation. They are briefly described below.

Forest Management Certification

This certification is awarded to forest managers after it is shown that their forestry operations meet or exceed the FSC criteria.

Chain-of-Custody Certification

The chain-of-command custody is awarded to companies that process, manufacture, and sell products made from certified wood. These companies are audited annually to assure that the materials from FSC forests are documented and kept separate from materials the company might use from non-FSC sources. If the product is a fiber-based product such as a composite wood material, the audit will ascertain that the products contain the minimum fiber content as required by the FSC. The following diagram illustrates the FSC COC diagram (see Figure 10-2).

Figure 10-2

FSC chain of custody diagram.

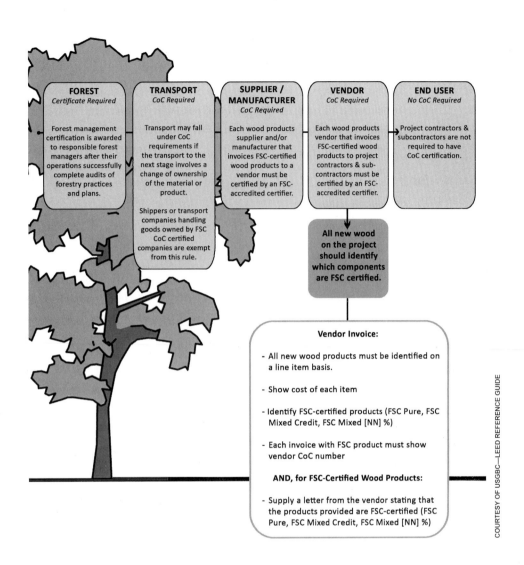

FOREST
Certificate Required

Forest management certification is awarded to responsible forest managers after their operations successfully complete audits of forestry practices and plans.

TRANSPORT
CoC Required

Transport may fall under CoC requirements if the transport to the next stage involves a change of ownership of the material or product.

Shippers or transport companies handling goods owned by FSC CoC certified companies are exempt from this rule.

SUPPLIER / MANUFACTURER
CoC Required

Each wood products supplier and/or manufacturer that invoices FSC-certified wood products to a vendor must be certified by an FSC-accredited certifier.

VENDOR
CoC Required

Each wood products vendor that invoices FSC-certified wood products to project contractors & sub-contractors must be certified by an FSC-accredited certifier.

END USER
No CoC Required

Project contractors & subcontractors are not required to have CoC certification.

All new wood on the project should identify which components are FSC certified.

Vendor Invoice:

- All new wood products must be identified on a line item basis.

- Show cost of each item

- Identify FSC-certified products (FSC Pure, FSC Mixed Credit, FSC Mixed [NN] %)

- Each invoice with FSC product must show vendor CoC number

AND, for FSC-Certified Wood Products:

- Supply a letter from the vendor stating that the products provided are FSC-certified (FSC Pure, FSC Mixed Credit, FSC Mixed [NN] %)

Once these products are certified then their use in the project can be documented and LEED points awarded.

INTENT OF THE LEED REQUIREMENT

The intent of this LEED credit requirement is to encourage environmentally responsible forest management. USGBC accomplishes this by awarding credit points to the projects that use FSC-certified wood products. The LEED credit which is listed as LEED MR-7 is an award to the projects that can document that 50% of all wood or wood-based products employed in the project is FSC-certified wood. An additional point in the Innovation in Design Category is awarded if the project can document that 95% of all wood in the project comes from FSC-certified sources. Only wood materials permanently installed in the project can be counted. Furniture can be included in this calculation only if it has been included in the calculations of other similar points (see Figure 10-3).

Credits for Use of Certified Wood	
Percentage of Certified Wood	**Credits**
50%	1
95%	2

© CENGAGE LEARNING 2012

Figure 10-3

LEED MR-7 credits.

STRATEGIES FOR SUCCESS

The first step in compliance with the LEED requirements for these credits is for the team to establish a goal for FSC-certified wood in the project. The first target goal would be that 50% of all wood incorporated into the project meet the requirements for FSC certification. The achieving of this goal will result in the award of one credit point. The second goal would be a target of 95% certified wood that would result in the award of one additional credit point in the Innovation in Design category.

One of the keys to achieving the points for this credit is for the contractor to contact wood material suppliers early in the construction process. It is very important to identify wood material suppliers who can supply the types and quantities of FSC-certified wood on the project. As the availability of this certified wood often varies over the duration of a project, it is in the contractor's best interest to buy out the certified wood early in the construction process. To do this, the contractor must have an adequately weather-protected and ventilated area to store the material in until it is ready to be incorporated into the project. This would require the contractor to locate and secure a warehouse facility nearby to store the certified wood unit that is needed on the project.

In addition to the availability question, the cost of this certified wood will be more than traditional wood. The contractors must take care to review the requirements prior or during the bidding process to make sure that the cost of certified wood has been included in the project cost proposal. If a warehouse is required, the cost of this warehouse must also be included.

In order to document that the wood is indeed FSC–certified, the contractors must document the COC. To do this, they must make sure that the invoice from each wood vendor contains the following information:

- Each wood product must be indicated on the invoice as a separate line item.
- Each FSC-certified wood product must be clearly identified.
- The dollar amount for each FSC certified.
- Finally the vendor's COC certificate number must appear on the invoice.

As stated above, to achieve one point the total dollar amount of FSC-certified wood must be at least 50% of the total value of wood incorporated into the project. To calculate this, the contractor can use the following formula (see Figure 10-4).

Figure 10-4

Formula to calculate percentage of certified wood.

© CENGAGE LEARNING 2012

$$\text{Percentage of certified wood material} = \frac{\text{FSC certified wood material value (\$)}}{\text{Total new wood materials cost (\$)}} \times 100$$

For instance, assume that a new office building has a total wood cost of $250,000. The contractor through careful product and supplier selection was able to obtain $145,000 in FSC-certified wood products. Using the formula above, the value of certified wood would equal 58% of the projects total cost of wood. This would result in the award of one credit point.

$$58\% \text{ certified wood meterial} = \frac{\$145,000}{\$250,000} \times 100$$

On most construction projects wood is not always a stand-alone material but is a part of a composite assembly. To determine the value of certified wood in the assembly, the contractor must first determine the portion of the assembly that is certified wood. This is done by calculating the weight of the certified wood in the assembly and comparing it with the total weight of the assembly. The following formula will be helpful in performing this calculation (see Figure 10-5).

Figure 10-5

Formula to calculate percentage of certified wood in an assembly.

© CENGAGE LEARNING 2012

$$\text{Assembly certified wood material value} = \frac{\text{Weight of FSC certified wood material in the assembly}}{\text{Weight of the assembly}} = \text{Assembly value (\$)}$$

For instance, assume that a wall component assembly weighs approximately 1500 lbs and its value is $1750.00. If the total weight of the FSC wood in the assembly is 750 lbs, the value of the FSC-certified wood would be $875.00 and can be calculated as follows:

$$\$875.00 \text{ (assembly certified wood meterial value)} = \frac{750 \text{ lbs}}{1500 \text{ lbs}} \times \$1750.00$$

This value would be added to the other calculated FSC-certified wood material values to determine the total value of FSC-certified wood in the project (see Figure 10-6).

The following example illustrates how this method is employed with built-in cabinetry.

Figure 10-6

Sample certified wood assembly calculation.

Project Name: The Summit Office Building			
Manufacturer: Acme Cabinetry			
Product: Employee Lounge Cabinetry			
Component	**Weight (lbs)**	**Wood Based Weight (lbs)**	**FSC Certified Wood Weight (lbs)**
Wheatboard shell	60	60	60
Veneer top surface	5	5	0
Other wood shell structure	7	5	2
Non wood content	452		
Total Weight	524	70	62
Percent of wood in the component (wood weight / total weight) 13.3%			
Percentage of FSC Certified Wood (certified wood weight / total weight) 11.8%			

© CENGAGE LEARNING 2012

If the value of the built-in cabinetry is $4500, then the value of the certified wood would be

$$\$4500 \times 10.8\% \text{ (the percentage of certified wood)} = \$531.00$$

LEED DOCUMENTATION

The first step in the LEED online documentation process requires that the contractors select all other related credits that are being attempted, which include the following:

- MRc3: Materials Reuse.
- MRc4: Recycled Content.
- MRc5: Regional Materials.
- MRc6: Rapidly Renewable Materials.
- MRc7: Certified Wood.

Next the contractors must check off the box indicating whether or not furniture is included in their calculations.

The contractors must then upload information on all new wood-based products used in the project. They must provide further information such as material cut sheets for a minimum of 20% of the claimed items. It is important that they maintain in their files documentation on 100% of the items claimed to be FSC-certified in the event that the submit documentation is found to be deficient.

The contractors must check off a box certifying that all wood-based products have been included in the table. In lieu of providing full documentation, they can choose the expedited option in which a registered architect or interior designer will certify compliance.

11

INDOOR AIR QUALITY

INTRODUCTION

Unlike our ancestors, who often spent much of their time outdoors, according to the United States EPA, Americans currently spend approximately 90% of their lives inside buildings.[1] It is true that a vast majority of us live, eat, sleep, and work inside. Those working outside are the exception, not the rule. As most of our lives are spent inside, the quality of the indoor environment has become increasingly more critical. The oil embargo of the early 1970s caused an energy crisis in the United States. This resulted in a movement to conserve energy by designing and constructing our nation's buildings to consume less energy. Part of this effort resulted in the creation of both residential and nonresidential buildings, which were increasingly tighter.

Where our buildings were once designed to capture and direct outside air through the building, our modern building designs began to shun the concept of using free flowing air throughout the building. The free flow of fresh air has been replaced by the idea that a totally sealed building is more energy efficient. Although certain energy efficiencies might be gained, there is a cost. That cost is the potential danger to human health of an insufficiently ventilated building.

Many if not most buildings are composed of nonnatural materials that have been developed over the past 50 years or so. Many of these materials are heavily laden with chemicals. These materials include paints, varnishes, sealants and adhesives, foam insulation, particleboard, OSB, wall board, carpet, and ceiling tile to name a few. Once incorporated into the buildings, these materials often give off toxic gasses during a natural process referred to as off-gassing. As our buildings are generally constructed "air tight," these gasses have no way of being expelled from the building. They therefore continue to build up pollution within the indoor environment until they reach toxic levels and the building's inhabitants start to become sick.

Scientists are now finding that except in our most polluted cities, the pollution levels of air inside our buildings can be as great as two to five times greater that the air outside the building. In rare cases, the pollution levels inside a building have been measured as high as 100 times greater than that just outside the building.[2] It is hard to believe that it might be healthier to walk next to a heavily trafficked street rather than working within some of the most polluted buildings. A World Health Organization Committee study indicates that as many as 30% of newly constructed and remodeled buildings might be making their inhabitants ill.[3] This phenomenon has been referred to as "sick building syndrome" (SBS). This syndrome can be caused by anything from chemicals in the indoor environment to the presence of mold and bacteria in the buildings air flow. SBS is not a single illness but is a generalized name for the cause of a group of illnesses that can be directly attributed to a building's indoor air environment. These illnesses that are referred to as building-related illnesses (BRI) include occupational asthma, hypersensitive pneumonia, community acquired pneumonia, and Legionnaires' disease.

TAKE NOTE

Americans spend 90% of their time in buildings. The air quality in some buildings is sometime 100 times worse than at street level.

OCCUPATIONAL ASTHMA

Occupational or work-related asthma is a medical condition that consists of a narrowing of the breathing passages. It occurs in both children and adults and is caused by inhaling airborne particles within a building's interior environment. These particles act as irritants and cause an allergic reaction. Symptoms include shortness of breath, coughing, wheezing, runny nose, watering eyes, and occasionally a feeling of tightness in the chest. In the diagnosis of occupational asthma, the person would have had no prior history of asthma or breathing difficulty. The onset of the condition would be either entirely or primarily attributed to conditions within the building environment. The symptoms of occupational asthma are reversible if the irritants are removed from the building's air stream. Occupationally aggravated asthma is similar but the diagnosis would be preceded by a history of asthma or breathing difficulties. The Center for Disease Control and Prevention estimates that 15%–23% of new asthma cases in the United States each year can be attributed to occupational environmental hazards (see Figure 11-1).[4]

HYPERSENSITIVE PNEUMONIA

Hypersensitive pneumonia is a disease in which a person's lung sacs (alveoli) become inflamed. This inflammation of the lungs can eventually lead to the formation of scar tissue. The primary symptom of hypersensitive pneumonia is breathlessness and extreme breathing difficulties. Hypersensitive pneumonia is caused by a person being exposed to contaminated dust within a building's environment. If the dust contains certain microorganisms, proteins, or chemicals, the inhaling of the dust can cause severe allergic reactions in some people. Air-conditioner lung is a type of hypersensitive pneumonia caused by building air conditioning (AC) systems circulating air contaminated with antigens capable of causing these severe

Figure 11-1

Asthma sufferer.

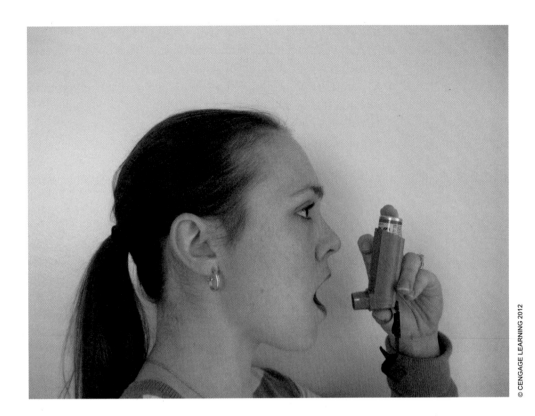

© CENGAGE LEARNING 2012

allergic reactions. The symptoms of this disease can be rapid onset with cough, shortness of breath, fever, and chills occurring within 4–8 h. after exposure or the symptoms can develop more slowly over several days or weeks.

LEGIONNAIRES' DISEASE

Legionnaires' disease is a special type of pneumonia caused by the *Legionella* bacteria. The bacteria and disease got its name in 1976 when a large group of people attending an American Legion convention in Philadelphia contracted a mysterious and deadly type of pneumonia. Although it appears clear that the disease existed before 1976, it is now being more commonly diagnosed. In fact, there are as many as 18,000 people hospitalized each year with this disease. Symptoms of Legionnaires' disease include a high fever, chills, and a cough.[5] The symptoms generally appear 2–14 days after exposure to the bacteria. The disease can be very serious depending on the age of the person with 5%–30% of the diagnosed cases resulting in death.

MOLD

Another cause of SBS is the presence of mold inside a building. Molds are naturally occurring organisms that can be found both inside and outside the buildings. These organisms multiply by releasing spores into the air stream. These mold spores are carried wherever the air stream takes them. Mold is generally not a great problem unless the spores land on a moist surface. Mold cannot grow on a dry surface but once in contact with a moist building material like wood or drywall, the mold can cause significant damage to both the building material and the health of the building's inhabitants (see Figure 11-2).

There are hundreds of thousands of types of molds, many of which have still not been totally identified. These molds can be roughly divided into three groups according to how they might affect a person's health. They are allergenic molds, pathogenic molds, and toxic molds.

Figure 11-2

Mold.

COURTESY OF WWW.SPACEHAGGIS.COM

Allergenic Molds

The first group is called allergic molds. These molds do not usually produce any life-threatening effects and are relatively safe in small amounts. They are generally more problematic to people with existing allergic conditions or to people suffering with breathing difficulty. The most common symptoms of allergic molds include scratchy throats, watery eyes, nose irritations, and in some cases the presence of a rash. Common allergenic molds include pathogenic molds and toxigenic molds.

Pathogenic Molds

Pathogenic molds are generally responsible for creating some type of infection. Most healthy people have strong enough immune systems to ward off any infections from most pathogenic molds. However, people with reduced or impaired immune systems such as those with HIV/AIDS, autoimmune diseases, and those undergoing chemotherapy are particularly susceptible to infection from exposure to pathogenic molds. Pathogenic molds would include the following:

Toxigenic Molds

The third group of molds called toxigenic molds is by far the most dangerous to human health. They are so toxic that exposure to them can endanger anyone's health regardless of their lack of previous illness.

The following is a list of the molds most commonly found in the indoor environment.

- *Alternaria* sp.—*Alternaria* is an extremely common and widespread type of mold. There are currently 44 identified species of *Alternaria* mold but there are probably hundreds more. It can be found on carpets and any other horizontal surface within a building's interior. It often resides on the inside portion of a building's windowsill. It produces rather large spores that are easily inhaled into the respiratory system. It has been known to cause both chronic sinusitis and some forms of asthma. If not corrected within a reasonable time, continued contact with the mold can cause pulmonary emphysema.
- *Aspergillus* sp.—*Aspergillus* mold is another form of commonly encountered molds within a building's interior environment. There are over 150 known species of *Aspergillus* mold. All of the *Aspergillus* molds are considered allergenic. This mold has been known to cause symptoms ranging from infections of the eyes and ears to severe pulmonary infections. Many of these molds produce mycotoxins that have associates with a variety of diseases in humans. Toxin production varies from species to species with some toxins being considered carcinogenic to the human population.
- *Cladosporium* sp.—The *Cladosporium* mold is the most common mold found in buildings today. It thrives in the summer when significantly larger numbers of the mold are encountered. It is commonly found inside air-conditioning ducts. It has been found to be a common cause of extrinsic asthma with chronic cases possibly developing into pulmonary emphysema.
- *Mucor* sp.—*Mucor* is often found in soil, dead plant material, horse dung, fruits, and fruit juice. It is also found in leather, meat, dairy products, animal hair, and jute. Being a zygomycetes fungus it may be allergenic (skin and bronchial tests). This organism and other zygomycetes will grow rapidly on most fungal media. It may cause mucorosis in immune-compromised individuals. The sites of infection are the lung, nasal sinus, brain, eye, and skin. Infection may have multiple sites.
- *Penicillium* sp.—Aw (water activity) 0.78–0.88. A wide number of organisms have been placed in this genera. The identification of species is difficult. It is often found in aerosol samples. It is commonly found in soil, food, cellulose, grains, carpet, wallpaper, and in interior fiberglass duct insulation. It is also found in paint and compost piles. It may cause hypersensitivity pneumonitis and allergic alveolitis in susceptible individuals. It is reported to be allergenic (skin). Some species can produce mycotoxins. It is a common cause of

extrinsic asthma (immediate-type hypersensitivity: type I). Acute symptoms include edema and bronchiospasms; chronic cases may develop pulmonary emphysema.

- *Stachybotrys* sp.—*Stachybotrys* mold is one of the most toxic molds encountered within our buildings. Because of its dark color it has been commonly labeled as "black mold" and it can be found on a variety of common building materials including drywall. Warmer temperatures and a relative humidity above 55% are ideal conditions for this mold to produce mycotoxins. In fact, several of the strains of this mold produce a highly dangerous toxin that is poisonous when inhaled. Toxins produced by *Stachybotrys* mold induce symptoms including diarrhea, sore throat, and headaches.

The primary danger from most molds is the exposure from inhalation. Long before the mold can be seen or the moldy odor is present, the mold is giving off allergens and toxins. As mentioned earlier, some types of molds grow they give off mycotoxins. Scientists have identified over 200 different types of mycotoxins and there are many more yet to be identified.

TAKE NOTE

Indoor air quality hazards

- Hypersensitive pneumonia
- Legionnaires' disease
- Mold
- Chemicals

CHEMICALS

Organic chemicals, better known as VOCs, are the primary ingredient in everything from cleaning products to paints and varnishes. These chemicals are also widely used in the construction industry. Some of the most common VOCs used in the construction industry include formaldehyde, benzene, perchloroethylene, and many more.

Many of the materials incorporated into and within our nation's buildings contain significant amounts of these organic chemicals. Once manufactured, these materials begin to break down. As materials slowly deteriorate over time, toxic chemicals begin to be released into the building's indoor environment. This phenomenon referred to as "off gassing" fills the building's air stream with a chemical mist. The AC system serves effectively to distribute these chemical vapors throughout the building. A study conducted by the EPA on indoor air pollution found indoor

VOC levels 10 times that of the outside air.[3] Some of the symptoms of VOC-related illness include the following (see Figure 11-3):

- conjunctival irritation, commonly referred to as redness of the eye
- nose and throat discomfort
- headache
- allergic skin reaction
- dyspnea, or shortness of breath
- declines in serum cholinesterase levels
- nausea and emesis, commonly referred to as vomiting
- epistaxis, commonly referred to as nosebleed
- fatigue
- dizziness

Figure 11-3

VOC-related illness symptoms.

Signs and Symptoms	Biological Pollutants	Volatile Organics	Heavy Metals	Sick Building Syndrome
Rhinitis, nasal congestion	Yes	Yes	No	Yes
Epistaxis	No	Yes	No	No
Pharyngitis, cough	Yes	Yes	No	Yes
Wheezing, worsening asthma	No	Yes	No	Yes
Dyspnea	Yes	No	No	Yes
Severe lung disease	No	No	No	Yes
Conjunctival irritation	Yes	Yes	No	Yes
Headache or dizziness	Yes	Yes	Yes	Yes
Lethargy, fatigue, malaise	Yes[5]	Yes	Yes	Yes
Nausea, vomiting, anorexia	Yes	Yes	Yes	No
Cognitive impairment, personality change	No	Yes	Yes	Yes
Rashes	Yes	Yes	Yes	No
Fever, chills	Yes	No	Yes	No
Tachycardia	No	No	Yes	No
Retinal hemorrhage	No	No	No	No
Myalgia	No	Yes[5]	No	Yes
Hearing loss	No	Yes	No	No

WORKERS' PRODUCTIVITY

Recent studies are showing that the quality of indoor air not only affects health but also can affect workers' performance. Poor indoor air quality is now seen as a major cause on not only illness as previously stated but it is seen as a significant factor in reduced workers' productivity. The National Energy Management Institute (NEMI) estimates that as much as $200 billion are lost each year due to the negative effects of poor indoor air quality on workers' production.[6] People working in buildings with healthier indoor environments have been found to be considerably more productive than people working in "sick buildings." The reduced productivity due to the fact that people tend to work slower and are therefore less productive when working in environments that have less fresh air or air that is tainted with chemicals. In addition, studies have also shown that people working in "sick buildings" tend to be sick more often than others working in cleaner buildings. The combination of the slower worker and the significant loss of working days due to illness results in a significantly lower rate of workers' production. This condition is similar in both industrial and office environments.

STRATEGIES FOR PREVENTION

The further chapters of the book will address specific strategies for the successful reduction or elimination of the materials that contribute to both BRI and the SBS. Each chapter will cover a separate class of building materials ranging from paints and adhesives to flooring and composite wood materials. It is important that the contractor understand the properties of every material that they incorporate into the building and the potential these materials have to harm human health.

More specifically the following is a discussion of strategies for addressing the hazards presented in Sections "LEGIONNAIRES' DISEASE" and "MOLD" above.

Legionnaires' Disease

The cause of Legionnaires' disease has been attributed to the *Legionella* bacteria developing with a building's AC system and being distributed throughout the building's air stream. Cooling towers and large evaporative condensers are often found to be the source of these bacteria. The water employed in these systems can serve as an ideal breeding ground for these bacteria. The following recommendations to reduce bacterial hazards in AC systems have been developed by the Department of Labor.[7]

Mold

Generally speaking most types of mold need water to survive and thrive. The contractor must act diligently to eliminate any sources of moisture penetrating into the building. All joints shall be carefully constructed and weatherproofed. In addition, all unnecessary sources of moisture within the building, during the construction process, must be eliminated. These are discussed in greater detail in Chapter 14.

Important design features include easy access or easily disassembled components to allow cleaning of internal parts including the packing (fill). The following features should be considered in the system design and installation:

- Enclosure of the system will prevent drift of water vapor.
- Design features that minimize the spray generated by these systems are desirable.
- System design should recognize the value of operating with low sump-water temperatures.
- Each sump should be equipped with a "bleed," and make-up water should be supplied to the sump.
- High-efficiency drift eliminators are essential for all cooling towers
  - Older systems can usually be retrofitted with high-efficiency models.
  - A well-designed and well-fitted drift eliminator can greatly reduce water loss and potential for exposure.

REFERENCES AND SOURCES

United States Environmental Protection Agency (EPA), http://epa.gov.

American Lung Association, http://www.lungusa.org.

National Institute for Occupational Safety and Health (NIOSH), http://cdc.gov/NIOSH.

National Safety Council, http://www.nsc.org.

Doctor Fungus, http://www.doctorfungus.org.

Medicine Net.com, http://www.medicinenet.com.

Environmental Health and Safety Online, http://www.ehso.com.

Global Health Center, http://www.ghchealth.com.

INDOOR AIR QUALITY MANAGEMENT

INTRODUCTION

The effect of the construction process on both the earth's atmosphere and a building's inhabitants has been clearly documented. USGBC, through the LEED certification program, has established goals for lessening the impact of construction on the quality of our air. Toward this end, LEED has established credit requirements in two separate categories. The first is defined in LEED Credit IEQ 3.1: Construction Indoor Air Quality Management Plan—During Construction, and it deals with the protection of the air quality during construction operations. The second is listed as LEED Credit IEQ 3.2: Construction Indoor Air Quality Management Plan—Prior to Occupancy. This credit addresses that precautions must be taken just prior to a building's occupancy. The following is a discussion of these two credits.

INDOOR AIR QUALITY MANAGEMENT DURING CONSTRUCTION

BACKGROUND

As most people know, the construction process is not a pristinely clean operation. The construction operation itself involves a number of operations that can cause potential danger to the surrounding atmosphere by contaminating the local atmosphere with construction dust. More importantly, actions taken during the construction process can actually affect final indoor atmosphere of the completed building. Materials ranging from dust to other harmful chemicals, if not carefully contained, can cause permanent damage to the building's indoor air environment. Some of these materials include the following:

1. *Dust*
   Dust is ever present in both our outdoor and indoor environments. Although it appears to the naked eye to be composed of small dirt-like particles, its actual composition is much more complex. Construction dust often contains large concentrations of Silica. Silica is a natural mineral commonly found in the soil. Beach sand, glass, and granite are all silica type materials. There are two primary types of silica materials.
   The first type of silica is called noncrystalline silica. This type of silica is found in glass, silicon carbide, and silicone. This type of silica dust does not pose much of a health risk.

The second type and most dangerous type of silica is called is crystalline silica. The most common form of crystalline silica is quartz. Quartz is a mineral commonly found in sand, gravel, clay, and a number of other types of rock. When these materials are crushed or sawed as is commonly found on construction sites, the resulting dust is heavily laden with silica dust. Breathing this dust has been found to be the cause of a serious lung disease called silicosis. Silicosis is a scarring and hardening of the lungs. Once the lungs are scarred, it becomes increasingly more difficult to breath. The damage caused by this silica dust is permanent and nonreversible. Silica dust has also been found to increase a person's risk of lung cancer. The amount of silica dust produced depends on several factors including, the material being handled and the types of tools used on that material. A study conducted by the University of Washington Department of Environmental and Occupational Health Sciences found that the amount of dust produced varied widely according to the equipment being used.[1] The following chart from that study illustrates the silica concentrations for different construction tools (see Figure 12-1).

Figure 12-1

Silica quartz concentrations.

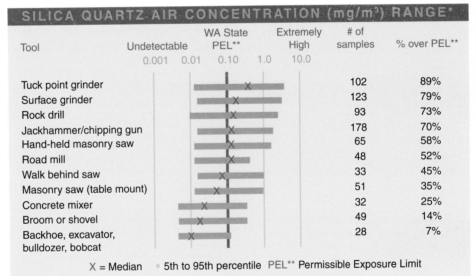

COURTESY OF UNIVERSITY OF WASHINGTON, DEPARTMENT OF ENVIRONMENTAL AND OCCUPATIONAL HEALTH SCIENCES, FIELD RESEARCH AND CONSULTATION GROUP

2. *Combustion Equipment*

There is a variety of combustion-driven construction equipment employed on a typical construction site. The number and size of this equipment varies with the phase of work under construction. The sitework phase generally uses the larger types of the equipment. Items ranging from small backhoes to much larger front-blade earth moving equipment are commonly employed during this phase of construction. The amount of exhaust released into the atmosphere from each piece of equipment will vary with the size and horsepower of that equipment. These emissions can greatly add to the level of green house emissions resulting from the construction process. The contractor must be very careful not to select equipment that exceeds the size necessary to effectively undertake the required construction operation. In addition, it is imperative that the equipment be in good working order and that it does not leak fluids or produce excessive exhaust emissions.

3. *Chemical Emissions*

As opposed to the two previously discussed air pollution issues, chemical emissions are generally more prevalent on jobsites during the actual building construction process. These are generally during the painting and finishing operations but they could occur at any time during the construction process. Chemical emissions, often referred to as VOCs, are emitted by a multitude of products used by the workers during construction. These products include the following:

- Paints
- Varnishes
- Sealants
- Adhesives
- Paint and varnish remover
- Other solvents

These VOCs are not only a hazard to our planet's atmosphere but they can be extremely hazardous to the building's inhabitants.

4. *Ozone Depleting Substances*

As mentioned earlier, our planet is surrounded by a protective layer of ozone. This ozone layer serves to protect our planet from the sun's harmful ultraviolet radiation. An ozone depleting substance is a chemical gas that rises to stratospheric elevations. Upon reaching the ozone layer, these chemicals are activated by the ultraviolet light to deplete the existing ozone (see Figure 12-2).

Figure 12-2

The ozone layer.

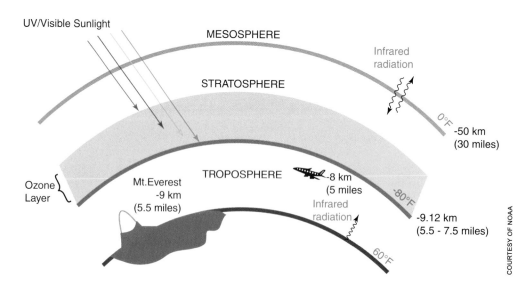

There are many ozone depleting substances that are in use in the construction industry. These include the following:

- Trichlorofluoromethane that are used in refrigerants.
- Dichlorodifluoromethane are also used in refrigerant and in older AC systems.

- Monochloropentafluoroethane is a gas used in the production of some types of insulation.
- Bromochlorodifluoromethane, which is commonly referred to as Halon, is used in both AC systems and fire suppression systems.

The following chart, from the EPA, is a list of the Class I ozone depleting substances (see Figure 12-3).

INTENT OF THE LEED REQUIREMENT FOR CONSTRUCTION INDOOR AIR QUALITY MANAGEMENT PLAN—DURING CONSTRUCTION

As mentioned earlier, indoor air quality (IAQ) is addressed in the LEED requirements in several different areas. First, the requirements in LEED Credit IEQ 3.1: Construction Indoor Air Quality Management Plan—During Construction addresses the protection of the building's indoor environment during the construction process (see Figure 12-4).

To qualify for the LEED Credit for Construction Indoor Air Quality Management Plan—During Construction, the contractor must develop and implement an IAQ Management Plan for the construction phase of the building.

- During construction, the contractor must meet or exceed the recommended control measures of the Sheet Metal and Air Conditioning National Contractors Association (SMACNA) IAQ Guidelines for Occupied Buildings under Construction, 1995, Chapter 3.

This SMACNA standard is the considered state-of-the-art in providing guidelines for protecting the indoor air environment of occupied buildings under construction from being adversely affected by the construction process.

The SMACNA standards address the following four components of IAQ control during construction.

1. *Protection of HVAC systems*
   The SMACNA guidelines require that all HVAC systems be protected from dust and debris during the construction operations. This includes all equipment and ductwork.
2. *Control of sources of pollution*
   The second component of the guidelines requires the contractor to address sources of pollution on the project site. This requires the contractor to consider VOC levels in all applied coating and finishes used on the interior of the building. It also addresses the need for storage of all VOC-emitting materials to be stored separately from any other building materials that might be able to absorb the VOC emissions.
3. *Pathway interruption*
   The guidelines discuss the need for contractors to isolate work within the building, which might be a source of dust or chemical emissions. By

Chemical Name	Lifetime, in years	ODP3 (WMO 2006)	ODP2 (40 CFR 82)	ODP1 (Montreal Protocol)	GWP5 (WMO 2006)	GWP4 (SROC)	GWP3 (40 CFR 82)	GWP2 (TAR)	GWP1 (WMO 2002)	CAS Number
Group I (from section 602 of the CAA)										
CFC-11 (CCl3F) Trichlorofluoromethane	45	1	1	1	4750	4680	4000	4600	4680	75-69-4
CFC-12 (CCl2F2) Dichlorodifluoromethane	100	1	1	1	10890	10720	8500	10600	10720	75-71-8
CFC-113 (C2F3Cl3) 1,1,2-Trichlorotrifluoroethane	85	1	0.8	0.8	6130	6030	5000	6000	6030	76-13-1
CFC-114 (C2F4Cl2) Dichlorotetrafluoroethane	300	1	1	1	10040		9300	9800	9880	76-14-2
CFC-115 (C2F5Cl) Monochloropentafluoroethane	1700	0.44	0.6	0.6	7370		9300	7200	7250	76-15-3
Group II (from section 602 of the CAA)										
Halon 1211 (CF2ClBr) Bromochlorodifluoromethane	16	7.1	3	3	1890	1860		1300	1860	353-59-3
Halon 1301 (CF3Br) Bromotrifluoromethane	65	16	10	10	7140	7030		6900	7030	75-63-8
Halon 2402 (C2F4Br2) Dibromotetrafluoroethane	20	11.5	6	6	1640	1620			1620	124-73-2
Group III (from section 602 of the CAA)										
CFC-13 (CF3Cl) Chlorotrifluoromethane	640		1	1	14420		11700	14000	14190	75-72-9
CFC-111 (C2FCl5) Pentachlorofluoroethane			1	1						354-56-3
CFC-112 (C2F2Cl4) Tetrachlorodifluoroethane			1	1						76-12-0
CFC-211 (C3FCl7) Heptachlorofluoropropane			1	1						422-78-6
CFC-212 (C3F2Cl6) Hexachlorodifluoropropane			1	1						3182-26-1
CFC-213 (C3F3Cl5) Pentachlorotrifluoropropane			1	1						2354-06-5
CFC-214 (C3F4Cl4) Tetrachlorotetrafluoropropane			1	1						29255-31-0
CFC-215 (C3F5Cl3) Trichloropentafluoropropane			1	1						4259-43-2
CFC-216 (C3F6Cl2) Dichlorohexafluoropropane			1	1						661-97-2
CFC-217 (C3F7Cl) Chloroheptafluoropropane			1	1						422-86-6
Group IV (from section 602 of the CAA)										
CCl4 Carbon tetrachloride	26	0.73	1.1	1.1	1400	1380	1400	1800	1380	56-23-5
Group V (from section 602 of the CAA)										
Methyl Chloroform (C2H3Cl3) 1,1,1-trichloroethane	5	0.12	0.1	0.1	146	144	110	140	144	71-55-6

Figure 12-3

Class I ozone depleting substances.

Figure 12-4

LEED IEQ 3.1 credit.

Credits for Construction Indoor Air Quality Management Plan—During Construction	
	Points
Develop and implement an Indoor Air Quality (IAQ) Management Plan	1

© CENGAGE LEARNING 2012

isolating these areas, the detrimental effects of the construction operations can be prevented from contaminating other areas of the building.

4. *Good Construction Housekeeping*

This component of the guidelines addresses the need for the contractor to maintain a clean and organized site. The contractor must keep the building's interior clean during the construction operations.

TAKE NOTE

Remember

A clean site is both a safe site and a healthy site.

5. *Scheduling*

This component of the guidelines addresses the need for the contractor to carefully schedule the installation of highly absorbent materials such as wall paper and carpets at times during the construction process so that the material's exposure to VOCs, other construction chemicals, and moisture are limited. For instance, carpeting and other highly absorbent materials should not be installed before painting. Once the carpeting has been installed, the use of any VOC-emitting materials must be limited.

STRATEGIES FOR SUCCESS

The first step in the compliance for this credit is for the contractors to prepare an IAQ Management plan. This plan must be completed prior to the commencement of the construction. The purpose of this plan will be to protect the HVAC system during construction, control pollutant sources, and interrupt contamination pathways. Because the quality of the indoor air during construction can and will have a significant impact on the final air quality of the building, the contractors should select an individual within their company to act as an IAQ compliance officer. This IAQ compliance officer will be responsible for overseeing air quality control procedures during construction.

The compliance officer should undertake a review of the construction plan and ascertain where and how the building's interior can be separated into isolated compartments. This will go a long way toward preventing the spread of construction contaminates. The materials for these construction barriers must be carefully selected and the barriers carefully constructed. The purpose is to stop the spread of contaminates such as construction dust and VOCs from one part of the building to another. These barriers can be constructed with any material that will be impervious to air flow. These include:

- Drywall
- Heavy gauge plastic sheeting (6 mil or greater)
- Closed cell foam, insulation

In addition to the plan, the contractor must carefully sequence the construction process. As stated earlier, the delivery timing, storage, and sequence of installation must be carefully planned. Sequence the installation of materials to avoid contamination of absorptive materials such as insulation, carpeting, and ceiling tile and gypsum wallboard. For instance, assume that the plumbing or other subcontractor is using adhesives for the installation of their work and that these adhesives give off harmful emissions. If there are porous materials such as gypsum wallboard stored in the same area, this wallboard will absorb the VOC emissions. Once the VOCs are in the wallboard, it will be very hard to get them out. Once permanently installed in the building, the contaminated wallboard will emit these VOCs back into the building's atmosphere for a long time.

Another area of the construction that must be carefully reviewed is the mechanical equipment. HVAC ductwork is often installed prior to the inter wall finishes such as gypsum wallboard. The cutting and finishing of this wallboard cause a large amount of dust to be released into the interior of the building. To prevent contamination of the duct system, sanding equipment with attached vacuums should be employed whenever possible. In addition, any opening within the duct system should be carefully sealed with plastic or some other material to prevent dust and other chemical particles to enter the duct system (see Figure 12-5).

At some point in time, the AC equipment must be started. If the construction operations required this AC startup to occur while there are still dusts or chemical operations ongoing, then the contractor should investigate using temporary AC equipment during this time. At the end, it could be less expensive to use temporary AC equipment rather than having to clean construction contaminates from the final equipment prior to occupancy.

If permanently installed air handlers are used during construction, equipment filters with a Minimum Efficiency Reporting Value (MERV) of 8 shall be used at each return air grille. In addition, these filters shall be replaced immediately before occupancy.

The diagram found on page 168, taken from the LEED User Web site, illustrates the steps necessary for compliance with the IEQc3.1 credit (see Figure 12-6).

Figure 12-5

Duct openings wrapped to prevent dust penetration.

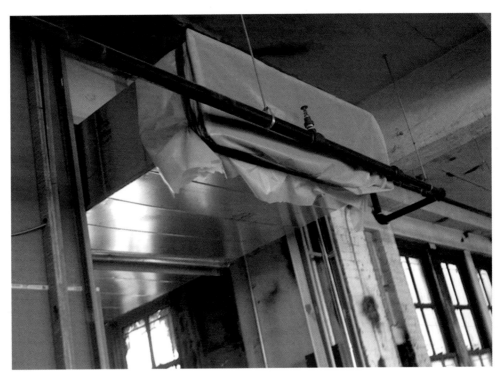

© CENGAGE LEARNING 2012

LEED DOCUMENTATION

The first step in the LEED online documentation requires that the contractor certify that an IAQ Management plan was developed and implemented during the construction. The next step in the documentation process is to upload the plan to the LEED online site. The documentation for this credit will require photographs indicating compliance to be submitted. Step three involves the writing of a narrative of how absorptive materials were protected from contamination. The last step of the process requires the contractor to certify that all the information submitted is in fact accurate.

The following is a sample LEED online template for this credit.

IAQ MANAGEMENT PRIOR TO OCCUPANCY

As mentioned earlier, the actual construction process is a potential source of contaminates within the building. Steps should be taken, by the contractor, to mitigate these contaminates during the construction operations. Recognizing that the contractor will never be 100% effective in eliminating the potential contamination hazard USGBC has established a requirement that the contractor undertake certain steps to cleanse the interior environment of the building after the completion of construction but prior to occupancy. The requirements for this cleansing is contained in LEED credit IEQc3.2 (see Figure 12-7).

Figure 12-6

IEQc3.1 LEED credit compliance diagram.

ACTION STEPS ✓

IEQc3.1: Construction Indoor Air Quality Management Plan-During Construction

POINTS: 1

Develop and implement an IAQ management plan for construction and pre-occupancy.

Integrate IAQ requirements into construction space

General contractor creates a comprehensive IAQ plan

Subcontractors oriented to the IAQ plan and their responsibilities

Avoid using building HVAC air handler during construction

OR

If HVAC used during construction: Install MERV-8 filters

Implement SMACNA 2007 IAQ plan guidelines

HVAC Protection: Cover ductwork, cover open grilles, etc.

Source Control: Use low-VOC Materials, limit indoor machine exhaust, use low-VOC cleaning products, etc.

Pathway Interruption: Use temporary barriers to separate construction from occupied areas and contain construction activities that produce lots of dust, exhaust area to create negative pressure, etc.

Housekeeping: Regular sweeping and wet mopping, etc.

Scheduling: Arrange construction schedule to limit occupant exposure to construction area, etc.

Document SMACNA activities with photos

Replace air filters if used during construction

COURTESY OF LEEDUSER.COM

Figure 12-7

LEED IEQ 3.2 credit.

Credits for Construction Indoor Air Quality Management Plan—Prior to Occupancy	
	Points
Perform Building Flush Out Operations or Provide Air Quality Testing	1

© CENGAGE LEARNING 2012

LEED REQUIREMENTS

In order to qualify for the IEQ 3.2 credit, the contractor can decide to take one of two options. The first option involves the undertaking of a total building flush out operation, whereas the second option involves the actual testing of the final indoor air environment.

FLUSH OUT OPTION 1/PATH 1

The first path in the total building flush out involves is employed after the completion of construction with all interior finishes installed but prior to occupancy. This operation involves the following two step process:

Step 1 Filtration Replacement

This step involves the replacement of all of the filtration media previously installed and used in the HVAC equipment. This replaced filtration media must meet or exceed a level of MERV 8.

Step 2 Building Flush Out

In order to comply with the requirements of this credit, the contractor must flush out the indoor environment of the building. In order to do this, the contractor must supply a volume of 14,000 cu. ft. of outdoor air per square foot of building floor area. In addition, during this process the indoor air temperature must be maintained at a minimum of 60 degrees with a humidity level not to exceed 60%.

FOR INSTANCE

Consider a business' corporate headquarters, which is a 27,700 sq. ft. office building, located in Chicago, Illinois. To comply with the LEED IEQc3.2, Step 2 flush out requirements, the contractor would have to supply fresh air as follows (see Figure 12-8).

Figure 12-8

Sample flush out air calculation Path 1.

Building Floor Area in sq. ft.		Flush out Air Requirement in cu. ft.		Total Required Flush Out Air in cu. ft.
27,700	×	14,000	=	387,800,000
Building Floor Area in sq. ft.		Flush out Air Requirement in cu. ft.		Total Required Flush Out Air in cu. ft.
27,700	×	14,000	=	387,800,000

© CENGAGE LEARNING 2012

Assume that the building has four AC units and that each will be bringing in outside air at a rate of 6000 cfm. That equates to a flush out air intake rate of 24,000 cfm. At that rate the flush out operation will take 11.5 days.

FLUSH OUT OPTION 1/PATH 2

The second path in the flush out option deals with a scenario when the building needs to be occupied prior to the completion of construction. The requirements for the total flush out is similar to path 1 but the rate of the flush out, because the building is occupied, is more controlled. In this path, once the building is ventilated, it must be ventilated with outside air at a rate of 0.30 cfm per sq. ft. of floor area. Using the building described earlier, the following figure illustrates the calculation of number of days for building flush out if the building is occupied (see Figure 12-9).

As the flush out rate is significantly lower than that used in path 1, the flush out period would extend over 32 days. This assumes that the AC equipment will be operating uninterrupted throughout a 24-hour day period.

Figure 12-9

Sample flush out air calculation Path 2.

Building Floor Area in sq. ft.		Flush out Air Requirement in cu. ft.		Total Required Flush Out Air in cu. ft.
27,700	×	14,000	=	387,800,000
Building Floor Area in sq. ft.		Flush out Air Requirement in cu. ft.		Total Required Flush Out Air in cu. ft.
27,700	×	14,000	=	387,800,000
Building Floor Area in sq. ft.		Flush out Air Requirement in cu. ft.		Total Required Flush Out Air in cu. ft.
27,700	×	14,000	=	387,800,000

© CENGAGE LEARNING 2012

OPTION 2 AIR TESTING

If for whatever reason the contractor or the LEED compliance team decided that the flush out process is not possible, there is a second compliance option in the LEED standard. This compliance method allows for the conducting of air quality testing. The testing procedure is established to be as follows:

1. The testing shall be undertaken after the completion of the construction but prior to occupancy.
2. All interior finishes must be completely installed. These include wallboard, paint, tile, carpet, millwork, and doors. Movable furnishings should be installed but are not required to be in the building at the time of the testing.

3. The AC equipment shall be operated during normally anticipated business hours beginning 3 h. prior to the anticipated opening of business. The rate of outside air shall be the same as that for the completed building.

4. The number of testing locations will be dependent on the size and design of the building. However, test should be undertaken at a minimum at the following locations:
   - One location for each 25,000 sq. ft. of floor area.
   - One location for each space supplied by a separate ventilation system.

5. The air samples shall be taken at an elevation of 3–6 ft. above the floor. These tests shall be taken over a 4-hour period.

The air tested shall meet or exceed the following standards (see Figure 12-10).

Figure 12-10

Maximum contaminate concentration levels.

Contaminant	Maximum Allowable Concentration
Formaldahyde	27 ppb
Particulates (PM 10)	50 µg/m³
Total Volatile Organic Compounds (TVOCs)	500 µg/m³
4-Phenylcyclohexene (4-PCH)*	6.5 µg/m³
Carbon Monoxide (CO)	9 ppm and no greater than 2 ppm above outdoor levels
* This test is only required if carpets and fabrics with styrene butadiene rubber (SBR) latex backing are installed as part of the base building systems.	

LEED DOCUMENTATION

The first step in the submittal of online documentation for this credit involves the uploading of the IAQ Management plan. The contractors must then input the date of building occupancy and they must select the type of pre-occupancy air quality assurance provisions employed. They include the following three options:

- Pre-occupancy flush out
- Early occupancy flush out
- IAQ testing

REFERENCES AND SOURCES

ASHRAE Standard 62.1–2004 Chapter 6; (CIBSE) Applications Manual 10: 2005, Natural Ventilation in Non-Domestic Buildings; Carbon Trust Good Practice Guide 237.

LOW-EMITTING MATERIALS—ADHESIVES AND SEALANTS

13

BACKGROUND

Adhesives that are often called glue have been used in the construction industry for over 75 years. There are used to bond a variety of materials to one another including but not limited to wood, paper, metals, and drywall. The following are examples of where adhesives and are used in the construction industry (see Figure 13-1):

- Prefabricated bonding of wood to wood
  - Plywood
  - Laminated lumber
  - OSB
  - Microlam members
  - Particleboard
- Bonding of furring to concrete or masonry
- Installing ceramic tile on floor or walls
- Installing vinyl or linoleum floor tile
- Installing cove base

Figure 13-1

Construction adhesive application.

- Installing glazing in window frames
- Installing glued down carpet installation
- Welding PVC, CPVC, and EBS piping
- Installing plastic laminate
- Installing single-ply roofing membranes

It has been reported that there are over 250,000 different adhesive and sealant products on the market today. The choice of the category of adhesive depends on several factors including the specific characteristics of the materials to be joined, the environmental resistance required, and temperature and moisture conditions, to name a few. In sustainable construction, a major factor in the adhesive or sealant selection process is the levels of toxic fumes given off by the material. Therefore, the VOC level of the adhesive or sealant becomes a primary factor in the product's selection.

There are four different general categories of adhesives. These categories differ from one another in the mechanism of the bonding process each uses. These categories are as follows:

1. *Solvent-based adhesives*

   Solvent-based adhesives consist of a polymer dissolved into a solvent. This is one of the oldest forms of construction adhesive. The liquid mixture is applied to one or both surfaces intended to be bonded together. Once the adhesive has been applied, the materials are joined together. As the solvent evaporates, the remaining polymer forms a bond between the two materials and harmful VOCs are released into the air stream. Solvent-based adhesives emit the highest quantity of VOCs.

2. *Water-based adhesives*

   Water-based adhesives are similar to that of the solvent-based adhesives except that in this case the polymer is dissolved in water. The process of adhesion is the same as for the solvent-based adhesive in that as the water is absorbed the polymer particles develop the bond between the two materials to be connected. There are pros and cons on the use of water-based adhesives. First, as water-based adhesives do not use solvents, the emission of harmful VOCs are minimal. However, water-based adhesives are often not as strong as solvent-based adhesives. This is generally not a problem because while weaker they are usually strong enough to effective do the job. As the bonding process of both the solvent-based and water-based adhesives requires a liquid, either the solvent or the water to be absorbed at least one of the two materials to be bonded together must be porous.

3. *Hot melt adhesives*

   Unlike the previous two categories of adhesives, this adhesive's natural state is a solid. This solid-state glue is heated to approximately 150° at which time it turns into a liquid. The hot adhesive is then poured onto the cooler surfaces, which are then placed together. As the glue cools, it returns to its solid state forming a strong bond between the two materials. Again as heat is used in place of a solvent, hot melt adhesives generally emit a lower level of VOCs into the air stream.

4. *Reactive adhesives*

This category adhesive is by far the most sophisticated of the four categories. In this process, two different chemical compounds are placed on the two opposites sides of the material to be joined. Once in contact with one another, a chemical reaction takes place between the two bonding compounds. This reaction forms a strong chemical bond between the two materials. Some examples of this category of adhesives would include the following:

- Epoxy adhesives
- Super glues
- Silicone sealants
- Reactive polyurethane sealants

As there is a chemical action taking place to form the adhesive bond, reactive adhesives can sometime emit a considerable quantity of VOCs.

TAKE NOTE

Construction adhesives

VOC levels from low to high

- Water-based
- Hot melt
- Reactive

INTENT OF THE LEED REQUIREMENT

The intent of this LEED requirement like that of the credit discussed in the previous chapter is to reduce the quantity of contaminants within the indoor air environment. As mentioned earlier, many construction materials are responsible for releasing toxic chemicals, VOCs, into a building's interior air stream. These chemicals when inhaled by the building's inhabitants can be harmful to the occupant's health and well-being. This specific LEED credit deals with the adhesives and sealants used in the construction of the building. It is listed in the Indoor Environmental Quality section in credit IEQ-4.1.

LEED REQUIREMENTS FOR LOW-EMITTING MATERIALS—ADHESIVES AND SEALANTS

The LEED standard requires that all adhesives and sealants used in the construction of the interior of the building comply with one of the following standards. This does not include weatherproofing sealants and adhesives used exclusively on the exterior surfaces of the building.

Standard 1 SCAQMD

This standard promulgated by the South Coast Air Quality Management District (SCAQMD) limits the amount of VOC that an adhesive or sealant can contain. The standard, which is Rule #1168, was first effective July 1, 2005 and amended January 2007.

South Coast Air Quality Management District (SCAQMD) Rule #1168. VOC limits are listed in the table below and correspond to an effective date of July 1, 2005 and rule amendment date of January 7, 2005.

The following chart illustrates the SCAQMD limits for VOC emissions in grams of VOC per liter of material (g/L). The lower the numerical level, the safer the material is for the environment. The contractor must select products that comply with these limits in order to be eligible for the respective LEED credit (see Figures 13-2 and 13-3).

Figure 13-2

SCAQMD VOC limits for adhesives.

Product Type	VOC level (in g/L)
ARCHITECTURAL ADHESIVES	
Product Type	**VOC level (in g/L)**
Ceramic Tile	65
Contact Cement	80
Fiberglass	80
Metal to metal	30
Multipurpose construction	70
Outdoor carpet adhesives	150
Rubber floor	60
Single ply roof membrane adhesives	250
Structural glazing adhesives	100
Wood: flooring	100
Wood: structural member	14
Wood: all other	30
All other adhesives (indoor carpet, carpet pad, subfloor, VCT & asphalt, cove base, dry wall & panel and all other adhesives)	50
Substrate Specific Applications (For any adhesive or primer not regulated above)	
Metal to Metal	30
Plastic Foams	50
Porous Material (except wood)	50
Wood	30
Fiberglass	80
Specialty Applications	
Special purpose contact adhesive	250
Welding: ABS	325
Welding: CPVC	490
Welding: plastic cement	250
Welding: PVC	510

Figure 13-3

SCAQMD VOC limits for sealants.

Product Type	VOC level (in g/L)
SEALANTS	
Product Type	VOC level (in g/L)
Architectural	250
Marie Deck	760
Non membrane Roof	300
Roadway	250
Single-Ply Roof Membrane	450
Other Sealants	420
SEALANT PRIMERS	
Architectural non porous	250
Architectural porous	775
Adhesive primer for plastic	650
Modified bituminous	500
Marine deck	760
Other Sealant Primers	750

© CENGAGE LEARNING 2012

Standard 2 Green Seal

The second referenced standard in the LEED requirement is the Green Seal Standards for Commercial Adhesives GS-36 (see Figures 13-4 through 13-6).

Figure 13-4

Green seal VOC limits for adhesives.

Product Type	VOC level (in g/L) minus water
Abs Welding	400
Carpet Pad Installation	150
Ceramic Tile Installation	130
Contact Bond	250
Contact Bond Specialty Substrates	400
Cove Base Installation	150
CPVC Welding	490
Indoor Floor Covering Installation	150
Multipurpose Construction	200
Non Membrane Roof Installation / Repair	300
Other Plastic Cement Welding	510
Outdoor Furniture Covering Installation	250
PVC Welding	510
Rubber Floor Installation	150
Single Ply Roof Membrane	250
Structural Glazing	100
Perimeter Bonded Sheet Vinyl Flooring	660
Waterproof Resorcinol Glue	170
Wood Flooring Adhesive	150

© CENGAGE LEARNING 2012

Figure 13-5

Green seal VOC limits for adhesives application to substrate.

Product Type	VOC level (in g/L) minus water
Flexible Vinyl	250
Fiberglass	200
Metal to Metal	30
Porous Material	120
Plastic Foams	120
Rubber	250
Other Substrates	250

© CENGAGE LEARNING 2012

Figure 13-6

Green seal VOC limits for aerosol adhesives.

Product Type	VOC level
General Purpose Mist Spray	65% by weight
General Purpose Web Spray	55% by weight
Special Purpose Aerosol Adhesives	70% by weight

© CENGAGE LEARNING 2012

PRODUCT SELECTION

It will be the designer's responsibility to specify the required level of compliance with either the SCAQMD or Green Seal standard. The contractor must study the specification and fully understand the designer's requirements. The following is a typical adhesive specification:

ADHESVIES AND SEALANTS (EQC4.1)

PART 1- GENERAL

1.1 SUBMITTALS

A. LEED Submittals:

Product Data for credit EQc4.1: For each adhesive and sealant used in the building's interior including printed statement of VOC content.

PART 2 - PRODUCTS

2.2 MISCELLANEOUS MATERIALS

A. Adhesives for _____:

Provide the product that meets these performance requirements and meets the following criteria:

    1. VOC Content: VOC content at or below levels allowed by South Coast Air Quality Management District (SCAQMD) Rule #1168. Aerosol adhesives: VOC content at or below levels allowed by Green Seal Standard GC-36.

B. Sealants used as fillers for _____.

Provide the product that meets these performance requirements and meets the following criteria:

    1. VOC Content: VOC content at or below levels allowed by Bay Area Air Quality Management District Regulation 8, Rule 51.

Comply with the requirements of Section 01570, CONSTRUCTION WASTE MANAGEMENT, for removal and disposal of construction debris and waste.

Products

Titebond/GREENchoice

Franklin International is a company that has been in existence since 1935. It produces a wide variety of construction adhesives and sealants. In addition to its traditional products, the company markets a full line of sustainable adhesives and sealants under the GREENchoice label, www.titebond.com (see Figure 13-7).

Figure 13-7

Titebond construction adhesive.

COURTESY OF TITEBOND®

The company's adhesive and sealant products include the following:

- GREEN*choice*™ Heavy Duty Construction Adhesive
- GREEN*choice*™ Premium Polyurethane Adhesive
- GREEN*choice*™ Projects & Repair Construction Adhesive
- GREEN*choice*™ Professional Drywall Adhesive
- GREEN*choice*™ Weatherproof Subfloor Adhesive
- GREEN*choice*™ Professional Cove Base Adhesive
- GREEN*choice*™ Professional Radon Sealant
- GREEN*choice*™ Professional Acoustical Sealant

Figure 13-8

Roberts adhesive products.

COURTESY OF QEP CO., INC.

DAP

While the DAP company's roots go back to 1865, it has been manufacturing products for construction since the early 1900s. The company currently has expanded to include an array of Green adhesives and sealants (www.dap.com).

Roberts

The Roberts company has been manufacturing construction sealants since 1938. They offer a large assortment of construction adhesives ranging from construction adhesives to specialty flooring adhesives. Within their product line are a variety of "green" products for use on LEED projects (see Figure 13-8).

STRATEGIES FOR SUCCESS

As will be discussed in greater detail in Chapter 14, most of this work will be undertaken by subcontractors. These subcontractors can range from waterproofing contractors to cabinet installers. It is the general contractor's responsibility to assure that all sealant and adhesive materials brought onto the jobsite comply with the applicable standards for LEED credit. To accomplish this, the contractor must meet with all subcontractors whom might be using these materials. They must explain the required VOC maximum limits to each sub and receive guarantees of their support and compliance. It is also recommended that signs as previously discussed be posted in conspicuous locations. These signs will act as reminders to all subcontractor of the need for LEED VOC limit compliance.

At or near the completion of the construction, the contractor will have to go online and submit the required adhesive and sealant information. In order to complete this required online LEED documentation, the contractor must accurately document the types of coating used and their respective VOC levels. The following chart can be used to tract these materials (see Figure 13-9).

CONSTRUCTION ADHESIVE VOC LIMITS						
Type of Coating	Manufacturer	Product Name	VOC Level g/L	VOC Limits g/L	Reference Standard	Quantity

© CENGAGE LEARNING 2012

Figure 13-9

Construction adhesive table.

LEED ONLINE SUBMITTAL

Once the adhesive and sealant materials are properly documented, the information can be transferred to the LEED online Web site. The first step in this online documentation process is to check off that the VOC emission levels for adhesives and sealants meets the requirements of the SCAQMD Rule #1168, January 7, 2005. Next the contractor must check off that aerosol VOC emissions meet the requirements of the Green Seal Standard for Commercial Adhesives GS-36. After checking off these two boxes on the online form the contractor must upload a detailed summary of the approach to be used for meeting these requirements.

REFERENCES AND SOURCES

South Coast Air Quality Management District (SCAQMD) Rule #1168 of October 3, 2003, http://www.aqmd.gov/rules/reg

Architectural Coatings; Green Seal Standards for Commercial Adhesives GS-36, http://greenseal.org

SCAQMD Rule 1168, Adhesive & Sealant Applications, January 7, 2005 amendment, January 1, 2007 levels (g/l, less water and exempt compounds), http://www.aqmd.gov

ARCAT, http://www.arcat.com/divs/sec/sec09911, This site provides a list of LEED compliant sustainable sealants and adhesives.

LOW-EMITTING MATERIALS—PAINTS AND COATINGS

BACKGROUND

As discussed earlier, chemicals in the materials used to construct buildings are becoming increasingly more dangerous to human health. Historically, one of the most problematic materials is the paints and coatings used on building interiors. Traditional commercial paints can contain as many as 1500 different chemicals. These chemically laden paints like many other building materials give off toxic gas after application. This process is commonly referred to as "off gassing." Because of the increased tightness of the building envelope, these gasses cannot escape from the building. They continue to circulate through the interior air stream contaminating the air that we breathe.

INTENT OF THE LEED REQUIREMENT

The intent of LEED Environmental Quality Credit 4.2 is to reduce the quantity of indoor air contaminants that are odorous, irritating, and harmful to the comfort and well-being of installers and occupants.[1] The USGBC believes that the reduction in the amounts of chemicals used in building materials is one step toward improving indoor environments of our nation's buildings. LEED has chosen to limit the level of VOCs in paints and coating by requiring these materials to meet the following standards.

TAKE NOTE

Leed paint specifications

- Interior paint   GS-11
- Anti-corrosive coatings   GS-03
- Clear finishes   SCAQMD

LEED REQUIREMENTS FOR LOW-EMITTING MATERIALS—PAINTS AND COATINGS

The requirement for low-emitting paints and coatings is listed in the Indoor Environmental Quality section, credit IEQ-4.2. To fulfill this requirement, paints and coatings used on the interior of the building (defined as inside of the weatherproofing system and applied on-site) must meet certain regional or national environmental standards. The standards chosen have been developed by independent environmental quality organizations. There are three different painting and coating categories identified in the LEED requirements.

- Interior paints
- Anti-corrosive and anti-rust paints
- Clear wood finishes and floor coatings

Each category has its own specific VOC emission limits which cannot be exceeded.

Interior Paint

The standard chosen by USGBC, for interior paint, has been developed by Green Seal. Green seal was founded in 1989 as an independent nonprofit organization dedicated to developing science-based environmental certification standards. Through its standard setting, certification, and education programs, Green Seal reviews, evaluates, and identifies products that are designed and manufactured in a way to minimize the product's impact on the environment. It also offers a scientific analysis that is distributed to consumers so that they might make more educated purchasing decisions regarding the environmental impacts of the products they buy. A third function of Green Seal is to ensure consumers that any product purchased bearing the Green Seal Certification Mark is in fact an environmentally friendly product. Lastly, Green Seal endeavors to encourage manufacturers to develop new products that will have less of an impact on the environment than those manufactured previously. The intent of Green Seal's environmental requirements is to reduce, to the extent technologically and economically feasible, the environmental impacts associated with the manufacture, use, and disposal of products.[2]

The organization introduced its first environmental standard in 1991. This standard was closely followed by the organization's first product certification in 1992. Currently, Green Seal offers eight environmental certifications to products used in the construction industry. These range from paint standards to standards for electric chillers.

The USGBC has selected the Green Seal GS-11 Standard as the minimum standard for paints and interior coatings. These would include: interior paints, anti-corrosive coatings, reflective coatings, and floor paint.

In order to meet the Green Seal standard, the product is restricted from containing any component that is a known carcinogen, mutagen, or a reproductive toxin. In addition, the product cannot contain any material that is a hazardous air pollutant or has been found to be ozone depleting. In addition, to meet the GS standard, the product is prohibited from containing certain known dangerous chemicals. These include the following:

- 1,2-dichlorobenzene, is an organic compound and derivative of benzene. It has been ranked as one of the most environmentally hazardous materials currently in use.
- Alkylphenol ethoxylates, which has been found to be toxic to the human endocrine system.
- Formaldehyde, a known carcinogen which has been attributed as the cause of numerous human illnesses. It also has been ranked as one of the most environmentally hazardous materials currently in use.
- Heavy metals, including lead, mercury, and cadmium.
- Phthalates, these are esters of phthalic acid which are primarily used as plasticizers.
- Triphenyltins and tributyltins, these chemicals again one of the worst to our environment are extremely toxic to both the human endocrine and immune systems.

The standard also severely restricts the level of VOCs that are contained in the product. VOCs are organic chemical compounds that are used as components in most traditionally manufactured paints. These chemicals are able to vaporize into the atmosphere after the paint has been applied. The VOC limits set forth by Green Seal GS-11 are measured in grams of VOC per liter of material. There are two separate standards, one for uncolored paint and one for colored paint that has the color added at the point-of-sale.

The following chart illustrates the limits for VOC emissions in grams of VOC per liter of material (g/L). The lower the numerical level, the safer the material is for the environment and the building's occupants. The contractor must select products that comply with these limits in order to be eligible for the respective LEED credit (see Figures 14-1 and 14-2).

Figure 14-1

GS-11 VOC limits. Excluding point of sale color.

Product Type	VOC level (in g/L)
Flat Topcoat	50
Non-Flat Topcoat	100
Primer or Undercoat	100
Floor Paint	100
Anti-Corrosive Coating	250
Reflective Wall Coating	50
Reflective Roof Coating	100

© CENGAGE LEARNING 2012

Figure 14-2

VOC limits with color added.

Product Type	VOC level (in g/L)
Flat Topcoat with colorant added at the point-of-sale	100
Non-Flat Topcoat with colorant added at the point-of-sale	150
Product Type VOC level (in g/L) Primer or Undercoat with colorant added at the point-of-sale	150
Floor Paint with colorant added at the point-of-sale	150
Anti-Corrosive Coating with colorant added at the point-of-sale	300
Reflective Wall Coating with colorant added at the point-of-sale	100
Reflective Roof Coating with colorant added at the point-of-sale	150

COURTESY OF GREENSEAL.ORG

Product Type	VOC level (in g/L)
Gloss	250
Semi-gloss	250
Flat	250

COURTESY OF GREENSEAL.ORG

Figure 14-3

GS-03 VOC limits for anti-corrosive coatings. Excluding point-of-sale color.

Product Type	VOC level (in g/L)
Flat	100
Non-Flat	50
Non-Flat High Gloss	150
Primer	100
Floor Paint	50
Anti-Corrosive Coating	100
Stains	250
Varnish	400
Waterproof Wood	100
Waterproof Concrete	100

© CENGAGE LEARNING 2012

Figure 14-4

SCAQMD VOC limits for clear finishes.

Anti-Corrosive Coatings

The standard chosen by the USGBC for anti-corrosive coatings used on interior metal components is Green Seal GS-03. The VOC limits for these coatings can be found on the Green Seal Web site at http://greenseal.org or in the table below (see Figure 14-3).

Clear Finishes

The agency selected by the USGBC to establish standards for clear wood finishes, floor coatings, stains, and shellacs is the South Coast Air Quality Management District (SCAQMD). The SCAQMD is a governmental agency that is responsible for air pollution control for all of Orange, Los Angeles, Riverside, and San Bernardino counties. As part of its operating responsibilities, the SCAQMD develops standards that are disseminated in the form of rulings. SCAQMD Rule 1113, Architectural Coatings is the standard chosen by the USGBC as the basis of the LEED credit for the clear finishes listed above. The limits set forth in the standard are contained in Figure 14-4.

MATERIAL OPTIONS

Generally speaking, the design professionals in LEED projects will be specifying no-VOC or low VOC paints and coatings. Contractors are often given either a detailed paint specification or a prescriptive standard that must be met. The following is an example of a typical Green Paint Specification.

When given a choice, the contractor must be careful in the final selection of these architectural coating materials. It will be the contractor's responsibility to assure that the products used in the project are in fact certified by the proper agencies as stated earlier, and they meet the specific LEED credit standard established by the USGBC. There are three general classifications of environmentally friendly paints and coatings; natural paints, Zero VOC paints, and low VOC paints.

Natural Paints

The first of these nontoxic paints would be the "natural paints." These paints are manufactured from all natural materials. These natural components include water, plant oils, plant dyes, essential minerals found in clay, chalk, and talcum. Other components include earth, natural latex, beeswax, and mineral dyes for color. Natural paints generally have a pleasant smell when applied and allergies and sensitivities to this type of paint have been very rare. Natural paints are by far the safest to both humans and the environment. The following is a list of some high quality natural paint manufacturers.

- The Real Milk Paint Company
  Milk paint is a totally natural paint made from milk protein with lime, clay, and earth pigments added for color. The paint is nontoxic and totally VOC free. It is manufactured and sold as a powder in 28 different colors. The painter can mix the required base colors with water to achieve a number of other desired colors. www.realmilkpaint.com
- Livos
  The Livos Company is a company that produces a complete line of organic stains and wood finishes. The coatings are manufactured from all natural ingredients including linseed oil, citrus oil, and natural waxes. The nontoxic low-VOC coatings are used primarily on wood surfaces. The stains and coatings are marketing under the Kaldet brand name. www.livos.com
- EcoDesign's BioShield
  The BioShield Paint Company was created in 1982 in New Mexico with the goal of manufacturing high quality natural paints and stains. Their paints and stains are composed of natural materials such as essential and seed oils, tree resins, beeswax, and natural pigments. The products are marketed under the BioShield product brand. www.bioshieldpaint.com
- Weather-Bos
  The Weather-Bos company manufactures an assortment of nontoxic, natural-based paints, stains, and coating for both interior and exterior use. Many of the company's finishes have up to 99% inert ingredients making them very safe for both humans and the environment. The products are marketed under the Weather-Bos product label. www.weatherbos.com
- SoyGuard
  This product is manufactured by the BioPreserve Company. It produces an assortment of exterior stains and sealers. It is perhaps best known for its SoyGurard wood sealer. This sealer is a tough long lasting wood preservatives made almost entirely from soy oil. www.soyguard.com
- Silacote
  Silacote USA manufactures a natural paint product made primarily from the earth's natural minerals like quartz. This nontoxic paint is available in 282 different colors and is highly impervious to sunlight and weather. This paint is primarily an exterior paint for use on concrete and other similar surfaces. www.silacote.com
- Anna Sova
  The Anna Sova Luxury Organics company produces a line of luxury organic paints. These paints are manufactured from 90% food ingredients. These include a milk paint base with other natural food-based pigments. The paint is available in hundreds of different colors, organized in fourteen palettes. www.annasova.com
- Green Planet Paints
  The Green Planet Paint Company manufactures a zero VOC paint made from plant resins and mineral pigments. The paint is available in two finishes, flat and eggshell and in up to 120 colors. www.greenplanetpaints.com

LOW-EMITTING MATERIALS: PAINTS AND COATINGS (LEED-NC EQc4.2)

PART 1—GENERAL

1.1 SUBMITTALS

A. LEED Submittals:

Product Data for credit EQc4.2: For each paint and coating used including a printed statement of VOC content and chemical components.

PART 2—PRODUCTS

2.2 MISCELLANEOUS MATERIALS

A. Paints, Coatings, and Primers for Interior Walls and Ceilings for _____"

Provide architectural paints, coatings and primers applied to interior walls and ceilings that meet these performance requirements _____ and meet the following criteria:

1. VOC Content: VOC content cannot exceed the VOC limits established in Green Seal Standard GS-11, Paints, First Edition, May 20, 1993 (Flats: 50 g/L—Non-Flats: 150 g/L).

B. Anti-Corrosive and Anti-Rust Paints Applied to Interior Ferrous Metal Substrates:

Provide anti-corrosive and anti-rust paints applied to interior ferrous metal substrates that meet these performance requirements _____ and meet the following criteria:

1. VOC Content: VOC content can not exceed the VOC limit established in Green Seal Standard GC-03, Anti-Corrosive Paints, Second Edition, January 7, 1997 (250 g/L).

C. Clear Wood Finishes, Floor Coatings, Stains, and Shellacs Applied to Interior Elements:

Provide clear wood finishes, floor coatings, stains, and shellacs applied to interior elements that meet these performance requirements _____ and meet the following criteria:

1. VOC Content: VOC content can not exceed the VOC limits established in South Coast Air Quality Management District (SCAQMD) Rule 1113, Architectural Coatings, rules in effect on January 1, 2004 (Clear wood finishes: varnish 350 g/L, lacquer 550 g/L—Floor coatings: 100 g/L—Shellacs: Clear 730 g/L, pigmented 550 g/L—Stains: 250 g/L)

At the end of each spec section, add:

"Comply with the requirements of Section 01570, CONSTRUCTION WASTE MANAGEMENT, for removal and disposal of construction debris and waste.

Zero VOC Paints

The second classification of architectural coatings is the "Zero VOC" paint. These are formulated to contain no VOCs. Zero VOC paints are manufactured by a number of companies including those listed below.

- Benjamin Moore

  The company manufactures three grades of eco-friendly paints. Eco Spec its most economical paint is available as a flat, egg shell, or semigloss finish in a

Figure 14-5

Benjamin Moore zero VOC paints.

PRODUCT #	DESCRIPTION	Actual VOC	LEED
	ZERO VOC PAINTS		
231	ECO SPEC INTERIOR PRIMER	0 g/l	X
223	ECO SPEC LATEX EGGSHELL	0 g/l	X
373	ECO SPEC WB INTERIOR LATEX FLAT	0 g/l	X
372	ECO SPEC WB INTERIOR LATEX PRIMER	0 g/l	X
374	ECO SPEC WB INTERIOR LATEX PRIMEREGGSHELL	0 g/l	X
376	ECO SPEC WB INTERIOR LATEX SEMI-GLOSS	0 g/l	X
219	ECO SPEC® LATEX FLAT	0 g/l	X
513	NATURA INTERIOR LATEX EGGSHELL FINISH	0 g/l	X
512	NATURA INTERIOR LATEX FLAT FINISH	0 g/l	X
511	NATURA INTERIOR LATEX PRIMER	0 g/l	X
514	NATURA INTERIOR LATEX SEMI-GLOSS FINISH	0 g/l	X

variety of colors. The Natura line is Benjamin Moore's middle quality paint. It is available in the same finishes as the Eco Spec line and is 100% free of VOCs. The top of the line paint manufactured by the company is marketed under the product name of "Aura." The following is a sample table indicating the LEED compliant paint available from Benjamin Moore (see Figure 14-5).

- Miller Paint
  The Miller Paint Company has been manufacturing paint since 1890. The company's latest line is marketed under the name "Arco Pure." This solvent-free paint has no VOCs and comes in the following four finishes flat, eggshell, satin, and semigloss. All Arco Purepaints contain a built-in antimicrobial protection to guard against mold and mildew. Additional information on Miller Paint can be found at www.millerpaint.com.
- Vista Paint Company
  The Vista Paint Company offers a "Carefree Earth Coat" line of paints. These paints are of low odor and contain zero VOCs. This 100% acrylic paint is offered in flat, velva-sheen, eggshell, and semigloss finishes. Each finish is available in 1320 different colors. Additional information on Vista Paint can be found at www.vistapaint.com.
- PPG Architectural Finishes
  The Pittsburg Paint Company, which was established in 1900, in addition to several other lines markets an environmentally sensitive paint line under the name "Pure Performance." This paint contains no VOCs and produces a very low odor. It is offered in flat, eggshell, and semigloss finishes. The line is available in any of the 1890 colors in the PPG Voice of Color system. Additional information on PPG Paint can be found at www.ppg.com.

- Timber OX Green

  The Timber OX company manufactures a line of bio-based stains and wood preservatives under the name "Timber OX Green." These stains are based on natural oils like castor oil as opposed to petroleum products. They are said to be environmentally safe and low VOC. The products are fast drying and can be used both on interior and exterior building surfaces. Additional information on Timber OX coatings can be found at www.timberoxgreen.com.

- Sherwin Williams

  The Sherwin Williams Paint Company has been manufacturing paints since the 1800s. In addition to their traditional paint line, they manufacture a low VOC line of paints under the "Harmony" product label. The Sherwin William's ecologically friendly stains are marketed under the "Duration Home" product label. Additional information on Sherwin Williams Paint can be found at www.sherwin-williams.com.

- Mythic Paints

  The Mythic Paint Company in Hattiesburg, Mississippi markets a line of environmentally friendly paints it calls the first high performance no VOC paint. The paint has very low odor and no toxic VOCs. Its extensive color palette includes 13 off whites, 54 pinks, 90 purples, and 144 blues.

- Ecoshield Paint

  Ecoshield paints are manufactured by the Dunn-Edwards Paint Company. The company manufactures a full line of Zero VOC paints in semigloss, low sheen, and flat finishes. In addition, the company has an Ecoshield primer and an Acopustikote Zero VOC flat ceiling paint. Additional information on Ecoshield paint can be found at www.dunnedwards.com.

- Yolo Colorhouse

  Yolo Colorhouse is an environmentally friendly paint company that was established in 2005. It manufactures a line of Zero VOC paints in a limited palette of 40 hues. The paint that is marketed under the Colorhouse brand is available in semigloss, eggshell, and flat finishes. Additional information on Yolo Colorhouse paint can be found at www.yolocolorhouse.com.

Low VOC Paints

The third classification of architectural coatings is the low VOC paints. These paints admittedly contain VOCs but these chemicals are present at significantly lower levels than in traditional paints. In order to qualify for LEED credit xx.x, the paint must have a VOC level of less than 50 g/L for flat paints and 100 g/L for non-flat paints. Some manufacturers of low VOC paints include the following: Miller Paint, Benjamin Moore, AFM, Dunn Edwards, Frazee, ICI-Devoe, ICI-Dulux, Kelly Moore, Pittsburgh Paints (PPG), Sherwin Williams, and Vista. There are tables from each manufacturer that show the actual VOC levels and if the specific product qualify for the LEED credit (see Figure 14-6).

PRODUCT #	DESCRIPTION	Actual VOC	LEED
532	AURA BATH AND SPA MATTE FINISH	44 g/L	X
524	AURA EGGSHELL	48 g/L	X
526	AURA SATIN FINISH	48 g/L	X
528	AURA SEMI-GLOSS	47 g/L	X
520	AURA® COLOR FOUNDATION	48 g/L	X
W626	BEN INTERIOR LATEX EGGSHELL	45 g/L	X
W624	BEN INTERIOR LATEX PRIMER	46 g/L	X
W627	BEN INTERIOR LATEX SEMI-GLOSS	44 g/L	X
W625	BEN INTERIOR LATEX FLAT	43 g/L	X
224	ECO SPEC LATEX SEMI-GLOSS ENAMEL	11 g/L	X
023	FRESH START® 100% ACRYLIC PRIMER	49 g/L	X
131	LATEX MULTI PURPOSE PRIMER/SEALERS	47 g/L	X
322	MOORE'S KITCHEN & BATH	93 g/L	X
0P04	SSHP ACRYLIC METAL PRIMER GS-03	52 g/L	X
386	STUDIO FINISHES TEXTURE SAND FINISH	36 g/L	X
285	SUPER CRAFT BLOCK FILLER	48 g/L	X
290	SUPER CRAFT LATEX FLAT	44 g/L	X
284	SUPER HIDE LATEX PRIMER UNDERCOATER	46 g/L	X
781	SUPER SPEC GREEN EGGSHELL FINISH	46 g/L	X
780	SUPER SPEC GREEN FLAT FINISH	45 g/L	X
785	SUPER SPEC GREEN LATEX PRIMER SEALER	44 g/L	X
782	SUPER SPEC GREEN SEMI-GLOSS FINISH	43 g/L	X
160	SUPER SPEC LATEX BLOCK FILLER	23 g/L	X
275	SUPER SPEC LATEX FLAT	44 g/L	X
270	SUPER SPEC PREP COAT HI BUILD LATEX PRIMER	13.6 g/L	X
172	SUPER SPEC SATIN FIL	42 g/L	X
153	SWEEP-UP LATEX FLAT	46 g/L	X
156	SWEEP-UP LATEX SEMI-GLOSS	28 g/L	X
508	WATERBORNE CEILING PAINT	34 g/L	X
532	AURA BATH AND SPA MATTE FINISH	44 g/L	X
524	AURA EGGSHELL	48 g/L	X
526	AURA SATIN FINISH	48 g/L	X
528	AURA SEMI-GLOSS	47 g/L	X
520	AURA® COLOR FOUNDATION	48 g/L	X
W626	BEN INTERIOR LATEX EGGSHELL	45 g/L	X
W624	BEN INTERIOR LATEX PRIMER	46 g/L	X
W627	BEN INTERIOR LATEX SEMI-GLOSS	44 g/L	X
W625	BEN INTERIOR LATEX FLAT	43 g/L	X
224	ECO SPEC LATEX SEMI-GLOSS ENAMEL	11 g/L	X

Figure 14-6 (*Continues*)

Benjamin Moore low VOC interior paint.

023	FRESH START® 100% ACRYLIC PRIMER	49 g/L	X
131	LATEX MULTI PURPOSE PRIMER/SEALERS	47 g/L	X
322	MOORE'S KITCHEN & BATH	93 g/L	X
0P04	SSHP ACRYLIC METAL PRIMER GS-03	52 g/L	X
386	STUDIO FINISHES TEXTURE SAND FINISH	36 g/L	X
285	SUPER CRAFT BLOCK FILLER	48 g/L	X
290	SUPER CRAFT LATEX FLAT	44 g/L	X
284	SUPER HIDE LATEX PRIMER UNDERCOATER	46 g/L	X
781	SUPER SPEC GREEN EGGSHELL FINISH	46 g/L	X
780	SUPER SPEC GREEN FLAT FINISH	45 g/L	X
785	SUPER SPEC GREEN LATEX PRIMER SEALER	44 g/L	X
782	SUPER SPEC GREEN SEMI-GLOSS FINISH	43 g/L	X
160	SUPER SPEC LATEX BLOCK FILLER	23 g/L	X
275	SUPER SPEC LATEX FLAT	44 g/L	X
270	SUPER SPEC PREP COAT HI BUILD LATEX PRIMER	13.6 g/L	X
172	SUPER SPEC SATIN FIL	42 g/L	X
153	SWEEP-UP LATEX FLAT	46 g/L	X
156	SWEEP-UP LATEX SEMI-GLOSS	28 g/L	X
508	WATERBORNE CEILING PAINT	34 g/L	X

Figure 14-6 (*Continued*)

Benjamin Moore low VOC interior paint.

STRATEGIES FOR SUCCESS

As most of the work contained in this category will be undertaken by sub contractors, it is imperative that all subs are onboard with the intent to comply. To assure this, the general contractor (GC) must at the start of the job have a meeting with each sub either individually or in small groups. The GC must fully define each sub contractor's responsibility for compliance and explain in detail how this compliance must be achieved. The need for careful product selection and accurate documentation is important. Another tool the contractor can use to reinforce compliance is to post signs on the jobsite. These signs will serve to remind all sub contractors that this is a LEED project site and that their work shall be governed accordingly. The following is a sample of a sign that can be posted in strategic locations on the site (see Figure 14-7).

During the course of construction, the GC must review all product submittals received from the sub contractors and material suppliers for compliance with the requirements for this credit.

In order to complete the required online LEED documentation, the contractor must accurately document the types of paints and coatings used and their respective VOC levels. The following chart can be used to tract these materials (see Figure 14-8).

REMINDER

This is a LEED Project

All sub-contractors must check all materials for VOC levels prior to applying these materials on site.

All paints, coatings and adhesives must comply with applicable VOC standards and be approved.

© CENGAGE LEARNING 2012

Figure 14-7
Sample warning sign.

TAKE NOTE

Remember

- Have frequent LEED compliance meetings
- Review product submittals for compliance
- Post reminder signs at strategic locations
- Record and document all product selections and VOC levels

LEED DOCUMENTATION

Once the paints and coatings materials are properly documented the information can be transferred to the LEED online Web site.

- Step 1 of this online documentation process involves the contractor checking off that the project emission levels from paints and coatings meets the requirements of Green Seal Standard GS-11, Paints, First Edition, May 20, 1993.
- Step 2 involves checking off that the VOC emissions from anti-corrosive and anti-rust paints comply with Green Seal Standard GS-03, Anti-Corrosive Paints, Second Edition, January 7, 1997.
- Step 3 involves checking off that the VOC emission levels from wood finishes, floor coatings, stains, shellacs, or primers meet the requirements of SCAQMD Rule #1113, Architectural Coatings, January 1, 2004.
- Step 4 involves uploading a comprehensive summary of how the credit requirements will be met to the LEED website.

PAINTS AND COATINGS VOC LIMITS						
Type of Coating	Manufacturer	Product Name	VOC level g/L	VOC limits g/L	Reference Standard	Quantity gals

© CENGAGE LEARNING 2012

Figure 14-8

Paints and coatings table.

REFERENCES AND SOURCES

Green Seal Standard GS-11, Paints, First Edition, May 20, 1993, http://www .greenseal.org

(SCAQMD) Rule 1113, Architectural Coatings, http://www.aqmd.gov/rules

ARCAT, http://www.arcat.com/divs/sec/sec09911. This site provides a list of LEED compliant sustainable paints.

California Paints, http://www.californiapaints.com/GreenPage/GreenProducts .html

LOW-EMITTING MATERIALS—FLOORING

BACKGROUND

Carpet

As discussed in Chapter 11, the pollution levels of indoor air in some buildings have been found to be many times more than that of the outside air. There are many reasons for the high levels of indoor air pollution. One of the primary reasons is that the carpeting installed in many of our nation's buildings give off VOC emissions long after the carpet's installation. This is primarily due to the considerable amount of chemicals used in the production of most carpets.

Originally, the carpeting material was taken directly from nature. In 1791, William Sprague opened the first carpet mill in the United States.[1] These early carpets were primarily made of cotton or wool. These natural materials were environmentally friendly not only because they were made from naturally occurring materials but also they contained no chemical materials that would give off dangerous gasses. Even the materials used to color these early fabric carpets were made of dyes produced from vegetables, flowers, and naturally occurring minerals.

Polyester was first discovered in the laboratories of the DuPont Company when it was found that alcohol and carboxyl acids when combined would create fibers. This discovery was soon followed by the creation of nylon and Dacron. All the three materials could be used to create a strong fiber that could be woven. In 1965, the first polyester-woven fiber carpet was introduced. The material caught on and the industry began producing millions of square yards of artificial fiber carpets for installation in our homes and offices.

Today the floors of many of our homes and office buildings are covered with carpeting. Most of these carpets are manufactured from artificial materials and chemicals. These chemicals include toluene, benzene, formaldehyde, ethyl benzene, styrene, and acetone. Many of these materials have been listed on the EPA's list of extremely hazardous substances. 4-phenyl cyclohexane or 4-PC, as it is commonly called, is the chemical that gives carpet the new carpet smell. It is used in almost all new carpets and it has been linked to a number of respiratory illnesses.

According to the Carpet and Rug Institute (CRI), most carpet fibers are composed of the following five materials.[2]

Nylon:
Two types, Nylon 6,6 and Nylon 6, are typically used in carpet. Nylon features include the ability to produce a variety of color schemes, luxurious soft "hand," excellent resiliency, durability, abrasion resistance, and texture retention. Other

features include good resistance to stains and soils. Nylon is considered to be suitable for all types of traffic. Solution dyed nylon provides excellent color fastness and increases stain cleanability.

Olefin (polypropylene):

Olefin features include its inherent and permanent resistance to water-based stains, its colorfastness, and resistance to static electricity. Olefin is resistant to moisture, making it the primary fiber used in indoor/outdoor carpet.

Polyester:

Polyester features include its luxurious soft "hand," inherent and permanent stain resistance, and ability to produce bright colors. Other features include good abrasion resistance, fade resistance, and cleanability.

Triexta:

Triexta features include its luxurious soft "hand," excellent colorfastness and bright colors, and inherent and permanent stain resistance. Other features include durability, resistance to fading, cleanability, texture retention, and resiliency.

Wool:

Wool is a natural fiber noted for its luxury and performance. Its features include soft "hand," high bulk, color variety, and inherent flame retardant characteristics.

Carpet Cushion

A carpet cushion is generally used under installed carpets and rugs. This cushion as the name implies adds a greater cushion effect making the carpet softer to walk on. Like the carpet discussed earlier, the early forms of this cushion were manufactured from natural materials. Cattle hair and later jute comprised much of the materials of the early carpet cushions. As the need for more cushion material increased, the available supply of these natural materials was incapable of meeting the need. The industry turned to man-made artificial materials. Currently, two of the most commonly used materials for carpet cushion is polyurethane foam and synthetic foam rubber. These materials are generally found in the form of prime foam or bonded foam. Prime foam is a first time use of the synthetic foam material, whereas bonded foam is comprised of recycled materials.

INTENT OF THE LEED REQUIREMENT

The intent of this LEED requirement that is listed as IEQ Credit 4.3 is to reduce the quantity of indoor air contaminants that are odorous, irritating, and harmful to the comfort and well-being of installers and occupants. The standard does this by limiting the allowable levels of emissions from all flooring materials.

LEED REQUIREMENTS FOR LOW-EMITTING MATERIALS—FLOORING

The LEED requirement for carpets and carpet padding is listed in the Indoor Environmental Quality section, credit IEQ-4.3. The standard offers two options for compliance;

Option 1

To comply with Option 1 of the LEED IEQ-4.3, all flooring must comply with all items in the following list that is applicable to the project.

1. *Carpeting*

   If the project has carpeting, then all carpet installed in the building interior must meet the testing and product requirements of the CRI Green Label Plus1 program.

   The CRI CGI Green Label program is a program of testing and certification for carpets and other carpet-related materials. The testing is performed by an independent laboratory that measures emission levels of the following 13 chemicals.

   - Acetaldehyde
   - Benzene
   - Caprolactam
   - 2-Ethylhexanoic acid
   - Formaldehyde
   - 1-Methyl-2-Pyrrolidinone
   - Naphthalene
   - Nonanal
   - Octanal
   - 4-Phenylcyclohexene
   - Styrene
   - Toluene
   - Vinyl acetate

   Emission levels are measured at 24 h. and 14 days. The following chart defines the maximum allowable emission levels for these chemicals. The maximum air concentration levels are measured in micro grams of emission per cubic meters of indoor air (see Figure 15-1).

2. *Carpet cushion*

   All cushion material placed under carpets installed in the building's interior must meet the requirements of the CRI Green Label program. As discussed above for caqrpets, the CRI's Green Label Program measures the level of emissions of VOCs from the carpet cushion. The testing program measures emissions from the following materials including the total level of VOCs, butylated hydroxytoluene (BHT), formaldehyde, and 4-phenylcyclohexine (see Figure 15-2).

Figure 15-1

CRI green label plus carpet program testing criteria.[3]

Carpet 24 Hour and 14 Day Emissions Test Criteria			
	24 Hour Testing		14 Day Not to Exceed
Target Contaminate	Maximum Emission Factor (EF) μgm²•hr	Maximum Air Concentration μg/m³	Maximum Emission Factor (EF) μgm²•hr
Acetaldehyde	130	70	130
Benzene	55	30	55
Caprolactam	130	70	190
2-Ethylhexanolic Acid	46	25	46
Formaldehyde	30	16.5	30
1-Methyl-2-pyrrolidinone	300	160	300
Napthalene	8.2	4.5	8.2
Noanal	24	13	24
Octanal	13	7.2	13
4-Phenylcyclohexene	50	27	50
Styrene	410	220	410
Toluene	280	150	280
Vinyl acetate	190	100	190

© CENGAGE LEARNING 2012

Maximum VOC Levels in Carpet Cushion	
Target Contaminate	Maximum Emission Factor (EF) μgm²•hr
BHT	300
Formaldehyde	50
4-PCH	50
TVOCs	1000

© CENGAGE LEARNING 2012

Figure 15-2

Maximum VOC levels for carpet cushion

3. *Carpet adhesive*

Carpets can be installed by using a tack down strip generally located along the perimeter of the room or through a direct glue down method. The tack down method does not increase or affect the level of emissions coming from the carpet. However, the glue down method has the potential of being the source of considerable volume of harmful emissions. To control the level of these added emissions, the LEED standard requires that all carpet adhesive must meet the requirements of IEQ Credit 4.1: Adhesives and Sealants, which is covered in Chapter 13. The maximum level of VOCs has been set at 50 g/L.

4. *Hard surface flooring*

All hard surface flooring must be certified as compliant with the FloorScore2 standard. The FloorScore standard has been developed by the Resilient Floor Covering Institute in conjunction with Scientific Certification Systems (SCS). The SCS has been tasked with testing and certifying hard surface flooring and flooring adhesive products. These products include vinyl, linoleum, laminate flooring, wood flooring, ceramic flooring, rubber flooring, and wall base. They are tested for strict compliance with established

Figure 15-3

Maximum VOC levels for hard surface flooring cushion

Maximum VOC Levels in Hard Surface Flooring Cushion	
Target Contaminate	**Maximum Emission µg/m³**
Formaldehyde	16.5
Acetaldehyde	9
All other VOCs	See footnote

© CENGAGE LEARNING 2012

maximum VOC emission levels. The testing focuses primarily on two chemicals, formaldehyde and acetaldehyde. The maximum VOC emission limits are shown in the following table (see Figure 15-3). All other organic chemicals with established Chronic Reference Exposure Levels (CRELs)— less than or equal to ½ the CREL as listed in the latest edition of the Cal/EPA OEHHA list of chemicals with noncancer CRELs.

5. *Hard surface flooring–alternate compliance method*

An alternative compliance path using FloorScore is acceptable for credit achievement: 100% of the non-carpet finished flooring must be FloorScore-certified and must constitute at least 25% of the finished floor area. Examples of unfinished flooring include floors in mechanical rooms, electrical rooms, and elevator service rooms.

6. *Hard surface flooring finishes*

Unlike carpeting in which the carpet is the final floor surface, hard surface floors generally require some type of finished top coat. Hard surface floors such as concrete, wood, bamboo, and cork floor finishes are generally coated with a sealer, stain, or other applied finish. The LEED standard requires that all hard surface floor finishes meet the requirements of SCAQMD Rule 1113, Architectural Coatings, rules in effect on January 1, 2004. These requirements are discussed in Chapter 14.

7. *Tile setting adhesives*

Tile flooring materials such as ceramic, stone, granite, marble, and porcelain require the application of setting materials to bond the tile to the rough floor. These setting materials are usually cementations mortar or adhesive mastic-based. These setting materials can sometime be a source of potential chemical emissions into the building's air stream. The LEED standard requires that all tile setting adhesives and grout must meet SCAQMD Rule 1168. VOC limits correspond to an effective date of July 1, 2005 and rule amendment date of January 7, 2004. A more specific discussion of this requirement is contained in Chapter 14.

Option 2

The second option for compliance requires that all flooring elements installed in the building's interior must meet the testing and product requirements of the California Department of Health Services Standard Practice for the Testing of Volatile Organic Emissions from Various Sources Using Small-Scale Environmental Chambers, including 2004 Addenda.

STRATEGIES FOR SUCCESS

Like all of the LEED requirements for materials installed in the building, the contractors must establish control of these materials as soon as possible in the construction process. They must meet with each material supplier or subcontractor and clearly define the emission limits for the various materials each will be supplying or installing. The general contractor must develop a procedure for receiving, reviewing, and approving all material submittals for LEED compliance. Careful attention must be paid to LEED emission compliance standards as well as all other architectural requirements established for the materials by the design team. Only 20% of the flooring compliance documentation, selected at random, must be uploaded to the LEED online Web site. The contractor must retain documentation on all flooring for use in the event that one or more of the submittals is found to be unacceptable. The following is a sample of a carpet specification sheet indicating LEED compliance with the Green Label Plus standard (see Figure 15-4).

Figure 15-4

Sample carpet specification sheet.

PacifiCrest

BATON ROUGE (P0740)

Machine	Cut Pile
Yarn System	XTI® type 6,6 nylon
Gauge	1/10
Yarn Construction	2 Ply Heatset
Stitches Per Inch	11.67
Finished Average Pile Height	.281"
Tufted Yarn Weight	36 oz. per sq. yd.
Approximate Total Weight	73 oz. per sq. yd.
Dyeing Process	Piece Dyed
Primary Backing	Woven Polypropylene
Secondary Backing	ActionBac®
Static Control	Built-in antistatic, warranted not to exceed 3.5kv
Soil Retardant	ROYALGUARD®
Density Factor (D)	4,612
Weight Density (WD)	166,035
Carpet Width	12'
Flammability Test	Passes Methenamine Pill Test DOC FF1-70
	Passes Critical Radiant Flux ASTM E-648 – Class I
	Passes Smoke Chamber Test ASTM E-662
CRI Green Label +Plus Indoor Air Quality #	GLP1360
Warranty	10-Year Limited Abrasive Wear, Limited Lifetime Antistatic
RN 56412 MADE in U.S.A.	Yarn may be substituted due to industry fiber shortages.

PacifiCrest Mills is a clean air partner. This carpet is tufted and dyed in our state of the art, low environment impact, award-winning facility in Irvine, California. Contact PacifiCrest or your representative for details.

Environmentally Preferable Products are certified by Scientific Certification Systems as having a lesser or reduced effect on health and the environment when compared with competing products that serve the same purpose.

LEED CREDIT LEED-CI v2.0	CATEGORY DESCRIPTION	QUALIFYING DESCRIPTION	PTS
EQ CREDIT 4.3 Environmental Quality	Low Emitting Materials Carpet	Low VOC Approved by CRI Green Label +Plus	1
MR CREDIT 5.1 Material Resource	Regional Materials 20% Manufactured Locally	100% of material cost for projects within 500 miles of Irvine, CA	1

Rev. 2/22/11

REFERENCES AND SOURCES

Carpets

The following is a listing of carpet manufacturers who offer one or more Green Label Plus carpet products.[6]

Atlas Carpet Mills, Inc.
City of Commerce, CA
800-367-8188
www.atlascarpetmills.com

Beaulieu of America
Dalton, GA
800-227-7211
www.beaulieugroup.com

Beijing Kingtop Carpet Co. Ltd.
Beijing, International/Overseas
010-80521884
www.ktcarpet.cn/english/news_info
.asp?xw
New

Bentley Prince Street
City of Industry, CA
626-934-2153
www.bentleyprincestreet.com

Bloomsburg Carpet Industries
Bloomsburg, PA
212-688-7447
www.bloomsburgcarpet.com

Blue Ridge Industries
Ellijay, GA
800-241-5945
www.blueridgecarpet.com

Burtco Enterprises, Inc.
Dalton, GA
800-241-4019
www.burtcocarpet.com

Camelot Carpet Mills
Irvine, CA
800-854-3258
www.camelotcarpetmills.com

Carpet Maker (Thailand) Co. Ltd.
BanPai, International/Overseas
+(66)-43-286734
www.carpetmaker.co.th/

Carpets International Thailand Public
Company Limited
Bangkok, Thailand
66(0)2976-0123
www.carpetsinter.com

Changzhou Around Globe Carpet
Mfg. Co., Ltd.
Changzhou, Jiangsu Province, China
+86-0519-83292559
www.aroundglobe-carpet.com

Cherokee Carpet Industries
Dalton, GA
706-277-6277
www.southwindcarpet.com

Chilewich Sultan LLC
New York, NY
212-679-9204
www.chilewich.com/plynyl/

Clayton Miller Hospitality Carpets
Dalton, GA
877-261-6334
www.clayton-miller.com

Colin Campbell & Sons
Vancouver, British Columbia, Canada
800-667-5001
www.naturescarpet.com

Constantine
Dalton, GA
706-277-8533
www.constantine-carpet.com

Couristan Carpets, Inc.
Fort Lee, NJ
800-223-6186
www.couristan.com

CW Hospitality Weavers
Las Vegas, NV
702-247-4858
www.cambridgeweavers.com/

Dalian Jiamei Carpet Co.
Dalian, International/Overseas
+86-411-86732229

Desso
Waalwyk, The Netherlands
+31-(0)416-684-100
http://www.desso.com/Desso/
BusinessCarpets/bc_en

Dixie Home
Dalton, GA
800-273-8546
www.dixie-home.com

Dongsheng Carpet Group
Rizhao City, Shandong Prov., China
+86-633-8688999
www.dongsheng.com

Dura Tufting GmbH
Fulda, International/Overseas
+49-661/82 0
www.dura.de

Edel Tapijt B.T.
Genemuiden, International/Overseas
+31-(0)38-385-22-22
http://www.edel.nl/tapijt/index.php

Fabrica
Santa Ana, CA
800-854-0357
www.fabrica.com

Findeisen
Ettingen, International/Overseas
+49-(0)-7243 / 71000
www.finett.de/de/c1,0,0,0,0_Home
.html

Fortune Contract, Inc.
Dalton, GA
800-359-4508
www.fortunecarpet.com

Gulistan Carpet
Aberdeen, NC
800-869-2727
www.gulistan.com

Haima Group Corporation
Weihai City, Shandong Prov., China
+86-631-5201088
www.haimacarpet.com

InterfaceFLOR Commercial
LaGrange, GA
800-336-0225
www.interfaceflor.com

InterfaceFLOR Thailand
Panthong, Chonburi / Thailand
+66-38-214303
www.interfaceflor.com

J+J/Invision
Dalton, GA
800-241-4586
www.jj-invision.com

J. Mish Inc.
Cartersville, GA
678-605-9191
www.jmishinc.com

Jiangxi Huateng
Xinyu, International/Overseas
+86-790-6861389

Joy Carpets, Inc.
Fort Oglethrope, GA
800-866-2728
www.joycarpets.com

Julie Industries, Inc.
North Reading, MA
978-988-8802
www.staticsmart.com

Kaili Carpet Company, Ltd.
LiYang City, China
+86-519-7302785
www.kailicarpet.com

Kraus Carpet Mills Ltd.
Waterloo, ON, Canada
519-884-2310
www.krauscarpet.com

Lexmark Carpet Mills, Inc.
Dalton, GA
800-871-3211
www.lexmarkcarpet.com

Mannington Mills, Inc.
Mannington Commercial
Calhoun, GA
800-241-2262
www.mannington.com

Masland Carpets
A Dixie Group Company
Atmore, AL
800-633-0468
www.maslandcarpets.com

Mats, Inc.
Stoughton, MA
1-800-MATS-INC
www.matsinc.com

Miller Davis Group DBA Luzern
Limited
Chattanooga, TN
800-574-4790
www.luzernltd.com

Milliken and Company
LaGrange, GA
706-880-5511
www.millikencarpet.com

Milliken Textile (Zhangjiagang)
Co., Ltd.
Zhangjiagang, China, International/
Overseas
+86-21-6145-5555
www.millikencarpet.com

Moda Carpet
Irvine, CA
949-250-7123
www.modacarpets.com

Mohawk Industries, Inc.
Atlanta, GA
888-387-9881
www.mohawkind.com

Nood Fashion
Dalton, GA
888-351-0304
www.noodfashion.com

Northwest Carpets
Dalton, GA
800-367-2508
www.northwestcarpets.net

Nourison Rug Corporation
Saddle Brook, NJ
201-368-6900
www.nourison.com

PacifiCrest Mills, Inc.
Irvine, GA
949-474-5343
www.pacificrest.com

Ploy Regen Carpet Manufacturing
Co. Ltd.
Qing Yuan, International/Overseas
+86-21-6374-1875
ployregen.com/

R. Griffith Textile Co. Ltd.
R. Griffith Hospitality Carpet
Rizhao, Shandong, China

Royalty Carpet Mills
Irvine, CA
800-854-8331
www.royaltycarpetmills.com

Sacco Carpets
New York, NY
212-226-4344
www.saccocarpet.com

Scott Group Custom Carpets
Grand Rapids, MI
616-954-3200
www.scottgroup.com

Shanghai Judong Tile Carpet Co.
Jiading, International/Overseas
+86-21-59563033
judongchina.web.05120512.net.cn/

Shaw Industries, Inc.
Dalton, GA
800-441-7429
www.shawfloors.com

Siam Carpets Manufacturing
Co. Ltd.
Pathumthanee, International/
Overseas
+(66)-2-977-2800
siamcarpets.co.th/

Signature Hospitality Carpet
Dalton, GA
706-270-5799
www.signaturecarpets.com

Stanton Carpet Corporation
Syosett, NY
800-452-4474
www.stantoncarpet.com

Suzhou Tuntex Fiber & Carpet Co. Ltd.
Taican, International/Overseas
021-62333488-33
www.tuntex-carpet.com/en/index.asp

Tai Ping Carpets Americas, Inc.
Calhoun, GA
706-625-8905
www.taipingcarpets.com

Tandus Flooring Inc.—Crossley
Carpet Mills Ltd.
Dalton, GA
800-248-2878
www.tandus.com

Tandus Flooring, Inc.—Asia
Suzhou, China
800-248-2878
www.tandus.com

Tandus Flooring, Inc.—C&A
Dalton, GA
800-248-2878
www.tandus.com

Tandus Flooring, Inc.—Monterey
Carpets
Dalton, GA
800-248-2878
www.tandus.com

Tapetes Sao Carlos
Sao Carlos, International/ Overseas
+55-(16)-3362-4000
www.tapetessaocarlos.com.br/

Terza, S.A., de C.V.
el Carmen, NL, Mexico
5281-8748-4982
www.terza.com

The Natural Carpet Company
Venice, CA
310-664-1420
www.naturalcarpetcompany.com

Ulster Carpets Ltd.
County Armagh, Northern Ireland
28 38334433
www.ulstercarpets.com

Vebe Floorcoverings
Genemuiden, International/Overseas
+31-(413)-47-68-31
www.condorcarpets.nl/
vebe_floorcoverings?la

VI Floor/Van Dijk Carpet
Hasselt, International/Overseas
+31(0)38-4778181
vifloor.nl

Vorwerk & Co.
Hameln Germany, International/
Overseas
+49-5151103406

Weihai Shanhua Carpet Group Co. Ltd.
Weihai, Shandong, China
+86-631-5233009 (F)
www.chinashanhua.com

Westbond Ltd.
Brampton, Barnsley, UK
+44(0)1582-876-161
www.westbond-carpets.com

Woolshire Carpet Mills, Inc.
Calhoun, GA
800-799-6657
www.woolshire.com

Zhejiang Artistic Carpet Mfg.
Co. Ltd.
Hangzhou, Zhejiang Prov., China
+86-571-86224904
www.zhemeicorp.com

Zhengzhou Huade Mutual Benefit
Carpet Co. Ltd.
Henan Province, P.R. China
+86-371-64319418

Carpet Cushion

The following is a listing of manufacturers who offer one or more types of Green Label Plus carpet cushion.[7]

Green Label Plus Carpet Cushion Manufacturers

Manufacturer	Type Cushions	Product Number
Carpenter Company	Bonded polyurethane-Continuous process (CC-3)	CC-902240
Chamlian Enterprises	Resinated Synthetic Fiber (CC-9)	CC-400971
Chamlian Enterprises	Resinated Synthetic Fiber (CC-9)	CC-400971
Columbia Foam	Bonded-Log & Peel Process (CC-2)	CC-904691
Dalton Foam Division NCFI	Bonded-Log & Peel Process (CC-2)	CC-902377
DomFoam/Valle Foam	Bonded-Log & Peel Process (CC-2)	CC-315252
Dunlop Flooring, Inc. AU	Bonded polyurethane continuous, CC-3	CC-837529

(Continued)

Manufacturer	Type Cushions	Product Number
Dura Undercushions	Rubber (CC-10)	CC-902245
Future Foam Co- Council BLuffs IA	Bonded polyurethane, log & peel, CC-2	CC-837215
FXI Foamex Innovations Operating Company	Bonded-Log & Peel Process (CC-2)	CC-789123
FXI Foamex Innovations Operating Company	Prime Polyurethane (CC-1)	CC-974135
Healthier Choice Flooring	Mechanically Frothed Polyurethane (CC-5)	CC-842786
Hickory Springs Mfg. Co.	Bonded-Continuous process (CC-3)	CC-567345
Leggett & Platt	Resinated, or coated, Synthetic Fiber (CC-9)	CC-880170
Leggett & Platt	Rubber (CC-10)	CC-694132
Leggett & Platt	Prime polyurethane (CC-1)	CC-975130
Leggett & Platt	Bonded Polyurethane-Continuous Process (CC-3)	CC-970850
Leggett & Platt	Bonded (CC-2)	CC-976112
Leggett-Platt/Sponge Cushion Div	Rubber Cushion (CC-10)	CC-973417
Mohawk Industries	Synthetic Fiber (CC-8)	CC-903417
Mohawk Industries	Bonded polyurethane-log and peel process (CC-2)	CC-525051
Mohawk Industries	Prime polyurethane (CC-1)	CC-590601
Reliance Carpet Cushion-Gardena, CA	Resinated Synthetic Fiber, CC-9	CC-307281
Shaw	Mechanically Frothed Polyurethane CC-5	CC-968421
Shaw Industries	Synthetic Fibers (CC-8)	CC-967132
Vantage Industries, LLC	PVC Foam (CC-12), Laminate underlay	CC-235469
Vantage Industries, LLC	PVC foam (CC-12), PVC Coated Fabric	CC-756841
Vantage Industries, LLC	ECO Series, PVC Coated Fabric CC-12	CC-846555
Vitafoam Products Canada	Bonded-Log & Peel (CC-2)	CC-295360

Carpet Adhesives

The following is a listing of manufacturers who offer one or more types of Green Label Plus carpet cushion.[8]

Advanced Adhesive Technologies
Dalton, GA
800-228-4583
www.aatglue.com

APAC
APAC
Dalton, GA
800-747-7095
www.apacadhesives.com

BASF Building Systems
Skakopee, MN
800-496-6067
www.basf.com

Beaulieu Commercial
Chatsworth, GA
800-451-1250 ext. 1262
www.beaulieucommercial.com

Blue Ridge Carpet
Ellijay, GA
800-354-8951

Bostik, Inc.
Middleton, MA
888-592-8558
www.bostik-us.com

Clark Flooring Solutions
A Division of Clark Continental, Inc.
Pittsburgh, PA
800-837-4583
www.clarkgroup.org

CMH/Space Flooring
Wadesboro, NC
800-342-8523
www.cmhspaceflooring.com

DriTac Flooring Products, LLC
Clifton, NJ
800-394-9310
dritac.com

Durkan Patterned Carpets
Dalton, GA
706-278-7037
www.durkanpatternedcarpets.com

ECORE International
Lancaster, PA
717-824-8224
www.ecoreintl.com

Flexco
Tuscumbia, AL
800-633-3151
www.flexcofloors.com

Forbo
Huntersville, NC
704-948-0800
www.forbo.com/default.aspx

Gerflor/G2i, Inc.
Atlanta, GA
800-727-7505
www.g2ii.com

Great Northern Associates
Buffalo, NY
800-933-9336
www.greatnorthern-inc.com

InterfaceFLOR
LaGrange, GA
800-336-0225
www.interfaceflor.com

Interlock Industries Inc.
Dalton, GA
706-517-8989

J+J/Invision
Dalton, GA
800-241-4586
www.jj-invision.com

Lees Carpets
Kennesaw, GA
800-523-5647
www.leescarpets.com

Mannington Mills, Inc.
Mannington Commercial
Calhoun, GA
800-241-2262
www.mannington.com

MAPEI
Laval, Quebec
800-704-7986
www.mapei.com

Mats, Inc.
Stoughton, MA
800-628-7462
www.matsinc.com

Metroflor
Boca Raton, FL
866-687-6357
www.metroflorusa.com

Milliken & Company
LaGrange, GA
800-824-2246
www.milliken.com

Mohawk Industries, Inc.
Dalton, GA
800-622-6227
www.mohawkind.com

Para-Chem
Simpsonville, SC
800-763-7272
www.parachem.com

Powerhold
unknown, GA
800-555-5555
www.powerhold.net

PS America
Dalton, GA
706-277-9767

Roberts Capitol, Inc.
A Div. of Q.E.P. Company
Dalton, GA
706-277-5294
www.capitoladhesives.com

Roberts Consolidated
A Div. of Q.E.P. Company
Boca Raton, FL
800-423-6545
www.robertsconsolidated.com

Roppe Corporation, USA
Fostoria, OH
800-537-9527
www.roppe.com

Safe Landings, Inc.
San Antonio, TX
800-509-8888
safelandings.com

Shaw Industries, Inc.
Dalton, GA
800-441-7429
www.shawfloors.com

Solar Compounds Corp.
Linden, NJ
908-862-2813
www.solarcompounds.com

Specialty Construction Brands
Aurora, IL
800-552-6225
www.tecspecialty.com

Specialty Construction Brands/
Chapco

Chapco
Aurora, IL
630-978-7766
www.chapco-adhesive.com

Swiff-Train Company
Corpus Cristi, TX
361-883-1706
www.swiff-train.com

Tandus Flooring Inc.
C&A Floorcoverings; Crossley Carpet
Mills, Ltd.; Monterey Carpets
Dalton, GA
706-281-2713
www.tandus.com

Tarkett Sports
Peachtree City, GA
888-364-6541
www.tarketsports.com
Teknoflor
Boca Raton, FL
866-687-6357
www.teknoflor.com

W.F. Taylor Co., Inc.
Fontana, CA
800-397-4583
www.wftaylor.com

W.W. Henry Company
Aliquippa, PA
800-232-4832
www.wwhenry.com

XL Brands
Calhoun, GA
706-625-0025
www.xlbrands.com

LEED DOCUMENTATION

The LEED online submission requirements for this credit involve the contractors uploading information on the VOC levels from both the carpeting and the flooring adhesives to the LEED online Web site. In addition, they must certify that all flooring materials and adhesives have been included in the table.

LOW-EMITTING MATERIALS—COMPOSITE WOOD AND AGRIFIBER PRODUCTS

BACKGROUND

For the last 50 years, the use of composite wood products, in the construction industry, has increased considerably. These materials are used to fabricate a multitude of construction- and building-related products ranging from structural beams to furniture. There are roughly three major categories of composite wood products currently being produced. They include the following:

1. *Veneer-based composite wood products*
   These products are generally composed of thin veneers of wood laid on top of one another and synthetic resin together. They include products such as plywood and laminated veneer lumber (see Figures 16-1 through 16-3).
   - Plywood
     Plywood is used primarily as a sheet product, whereas laminated veneer lumber is used to manufacture beams and joists.
   - Laminated veneer lumber
   - Laminated timber

   A second subcategory of this veneer-based composite wood product is composed not of veneer material but of larger pieces of wood materials. These

Figure 16-1

Plywood.

COURTESY OF BYSTANDER/WIKIPEDIA

COURTESY OF KLSKI/WIKIMEDIA COMMONS

Figure 16-2

Laminated veneer lumber.

COURTESY APA—THE ENGINEERED WOOD ASSOCIATION

Figure 16-3

Laminated timber.

are commonly referred to as laminated wood or laminated timber and are generally used as joists, beams, and arches.

2. *Particle-based composite wood products*

The second category of composite wood products is composed of smaller pieces of wood usually derived from wood scraps. These wood scraps are synthetic resin together and pressed into a wood composite panel material. These products generally include particleboard, medium density fiberboard, OSB, and parallel strand lumber (see Figure 16-4 through 16-7).

- Particleboard

    Particleboard is composed of small particles of wood that are pressed together into panels using a synthetic resin binder. These panels are then used for a variety of nonstructural uses such as door cores, inexpensive furniture, and some shelving and cabinetry. Figure 16-4 is an example of particleboard.

- MDF

    MDF is similar to particleboard except that it is composed of much small wood fibers. When compressed with a synthetic resin binder, the result is a dense composite wood panel material, which is commonly used for cabinetry and furniture. Figure 16-5 is an example of MDF.

- High density fiberboard

    High density fiberboard, which is commonly referred to as hardboard, has been used in the construction industry for many years. It is composed of exploded wood fibers that have been highly compressed. These compressed fibers are bonded together with a synthetic resin to form a strong, dense sheathing material. In addition to interior uses like peg board and furniture backing, it is used as an exterior siding material to simulate rear wood siding.

Figure 16-4

Particle board.

Figure 16-5

Medium density
fiberboard.

Figure 16-6

Oriented strand board.

COURTESY OF IDAC LOGIC

Figure 16-7

Parallel strand lumber.

COURTESY OF AMERICAN WOOD COUNCIL

- OSB

  OSB is a wood composite panel composed of wood chips in a synthetic resin binder. The wood chips are generally oriented such that the board has a directional characteristic similar to the grain of wood. Next to plywood, this is the most structurally capable of the composite wood panel products. Depending on the applicable building code, OSB can be used for installations from floor and roof sheathing to exterior wall sheathing. Figure 16-6 is an example of OSB.

- Parallel strand lumber

  Parallel strand lumber is lumber, that is dimensional type pieces of wood, which is composed of linear pieces of wood combined with a synthetic resin binder and compressed. This type of composite wood derives its name from the fact that the strips of wood are oriented longitudinally

Figure 16-8

Wood plastic component.

COURTESY WIKIPEDIA/VARUNRAJENDRAN

along the long axis of the wood member. Figure 16-7 is an example of parallel strand lumber.

3. Wood plastic composite

Wood plastic composite (WPC) is the third category of wood composite materials used in construction. Although WPC is a relatively new material for construction, its popularity is gaining. It is manufactured from almost 100% recycled wood and plastic waste products. Sawdust is mixed with pulverized plastic to form a paste that can be formed or extruded into a variety of shapes including panels and molding (see Figure 16-8).

IAQ AND WOOD COMPOSITE MATERIALS

As a building material, the wood composite products discussed above serve the construction industry well. They are used for everything from structure components to furniture. The problem with these materials lies in the materials from which they are manufactured. They are all composed of some type of natural wood material either in strands, pieces, or dust which is combined with a synthetic bonding resin. It is the synthetic bonding resins that have been found to cause the problems. The three most common types of synthetic resins used in the manufacturing of these wood composite materials are as follows:

- Urea formaldehyde resins

  Urea formaldehyde (UF) resins are the most common, least expensive of the resins used in composite wood. Immediately after the product containing the UF is manufactured, it immediately begins to release harmful VOCs onto the atmosphere. Once the product has been incorporated into the building, these VOCs are trapped in the building's interior air environment. Although it is true that formaldehyde gasses are present in both the indoor and outdoor environment, the levels of these gasses are generally not above .03 ppm. The EPA has determined that these lower levels do not present a considerable health risk to the public. By introducing these gasses into the closed indoor environment, levels significantly above the danger threshold can be achieved. These higher levels do present a considerable health risk to the building's occupants.

- Phenol formaldehyde resins

  Phenol formaldehyde (PF) resins are similar to the UF resins except in this case the formaldehyde is mixed with a phenol, rather than urea. The result is a thermosetting resin with increased water resistance. Because of its waterproof nature, PF resin is commonly used in exterior products. However, the same off gassing of harmful VOCs occur so if this material was ever used on the interior of a building, it would be extremely harmful to the building's occupants.

- Melamine formaldehyde resin
This type of synthetic resin is both heat resistant and waterproof. It is commonly used in the manufacturing of surface materials in the interior of buildings. Commonly uses include plastic laminates and countertops.

Because these resins produce VOCs, which as stated before are harmful to the building environment and its occupants, their use should be strictly limited.

INTENT OF THE LEED REQUIREMENT

The intent of this LEED credit which is listed as IEQ 4.4 is to reduce the quantity of indoor air contaminants that are odorous, irritating, and harmful to the comfort and well-being of installers and occupants. To do this the standard strictly limits VOC emissions for building materials and products that are used within the building envelope.

LEED REQUIREMENTS FOR LOW-EMITTING MATERIALS—COMPOSITE WOOD AND AGRIFIBER PRODUCTS

LEED Credit IEQ 4.4 states the "Composite wood and agrifiber products used on the interior of the building (defined as inside of the weatherproofing system) shall contain no added UF resins. Composite wood and agrifiber products are defined as: particleboard, MDF, plywood, wheat board, strawboard, panel substrates, and door cores. Materials considered fit-out, furniture, and equipment (FF&E) are not considered base building elements and are not included." One point is awarded to the project if the documentation is submitted indicating that the standard is met.

STRATEGIES FOR SUCCESS

In order to comply, the contractor must make sure that all wood or agrifiber composite materials installed in the building contain no UF resins. In addition, all field and shop applied adhesives must also contain not added UF resins. As stated in Chapters 3 and 4, the GC must make sure that all subcontractors and material suppliers understand the need for compliance with this standard and that they agree to comply. Each composite material and adhesive must be checked for compliance. Because of the complexity of this review process, the contractor should have a knowledgeable person assigned to this task.

TAKE NOTE

Note to contractor

Assign LEED product compliance documentation to a knowledgeable person within the company.

To do this, the contractor can employ the following four step process:

Step 1: Request product and materials submittals
The GC should request documentation from subcontractors and material supplier's product that these products and materials meet the standard. While a list of submittals is quite commonly defined by the contract documents this list might not always be broad enough to assure compliance with this standard. For instance, a millwork specification will rarely require the millwork subcontractors to document the UF levels of the products in their cabinetry. On a LEED compliant project this documentation is crucial.

Step 2: Review the documentation for compliance
Upon receipt and as expeditiously as possible, the GC must carefully review the submittals for compliance. Only products and material submittals that clearly document that there is no added UF can be approved.

Step 3: Review the actual products for compliance
Once the products or materials have arrived at the site the GC must check out each against the approved product or material submittal. Only products that match the approved submittal and are compliant can be allowed into the building. Products not in compliance must be refused.

Step 4: Submit the compliance documentation
The last step in this process is to submit the documentation into the LEED online form.

TAKE NOTE

Product compliance process
- Request product compliance submittal
- Review product compliance submittal
- Inspect product against submittal for compliance
- Submit compliance documentation to LEED online

MATERIAL OPTIONS

Because of the recently discovered dangers of installing products with added UF, there has been the development of several environmentally safe alternatives. Most of these products are derived from either renewable resources or composed of 100% recycled materials. Among these are the following:

- Wheatboard
  Wheatboard, which is also referred to as strawboard, is a building product made from wheat fibers, a completely renewable resource. Like particle board previously discussed, the wheat fibers are bonded together with a bonding agent and compressed into boards. However, unlike particle board and KDF, wheatboard contains no harmful resins and will produce no VOC emissions into the building's interior.
  One of the largest suppliers of wheatboard is the Keiri Corp, which is located in Chione. Their "wheatboard" is made from sorgum fibers. The following is a product submittal of this material (see Figure 16-9).
- Strawboard
  Masonite Primeboard is one of the largest manufacturers of strawboard.
  This environmentally safe hardboard is composed of straw fibers. It is used extensively in the manufacturing of doors.
- Environmentally safe particle board
  This type of particleboard, unlike the conventionally manufactured particleboard, is environmentally friendly because it is generally composed of 100% recycled material. In addition, the binder material has no UF. The uses of this board, which is FSC certified, are similar to other convention particleboards. Manufacturers of these boards include:
  - Evergreen Particleboard, Boise Cascade Corp.
  - Encore Board, SierraPine, Ltd.
  - Freeform Particleboard, The Collins Companies

Figure 16-9

Sample wheatboard product submittal.

Kirei & LEED

build green

LEED™ is a set of green building standards developed by the US Green Building Council. There are multiple categories with points available for various green building techniques and material usage. By accumulating points, projects can qualify for varying levels of LEED Certification. Points are scored by satisfying the requirements of each credit within a given category.

Kirei products are manufactured with rapidly renewable and recycled content, as well as low- or no-added-urea-formaldehyde adhesives, and can help projects gain LEED™ Green Building credit in the following areas:

Kirei Board

MR 4.1/4.2	**Recycled Content**
	Minimum 90% post-industrial recycled material: Sorghum Straw
MR 6	**Rapidly Renewable Materials**
	Manufactured with rapidly renewable raw material: Sorghum straw grown in yearly harvest cycle.
	Minimum 90% rapidly renewable material in Kirei Board:
EQ 4.4	**Low-Emitting Materials: Composite Wood**
	No added urea formaldehyde

Kirei Bamboo

MR 6	**Rapidly Renewable Materials**
	90% rapidly renewable material - Fast-growing Moso Bamboo
EQ 4.4	**Low-Emitting Materials: Composite Wood**
	Kirei Zero™ bamboo products utilize no-added-urea-formaldehyde adhesives that qualify for low-emitting material credits. If your project would benefit from this credit, please specify Kirei Zero™ bamboo panels or veneer.
*(MR 7)	**Certified Wood (Pending)**
	Kirei Bamboo products are manufactured using bamboo from sustainably managed forests and plantations. Kirei is in the process of attaining FSC certification for our entire bamboo panel supply chain. Once this certification is complete our Kirei Bamboo products will be eligible for inclusion in MR 7.

Kirei Wheatboard

MR 4.1/4.2	**Recycled Content**
	90% post-industrial recycled material: Wheat Straw
MR 6	**Rapidly Renewable Materials**
	Manufactured with rapidly renewable raw material: Wheat straw grown in yearly harvest cycle.
	90% rapidly renewable material in Kirei Wheatboard:
EQ 4.4	**Low-Emitting Materials: Composite Wood**
	No added urea formaldehyde

Kirei Coco Tiles

MR 4.1/4.2	**Recycled Content**
	30-40% post-industrial recycled material: Coconut Shells
MR 6	**Rapidly Renewable Materials**
	Manufactured with rapidly renewable raw material: Coconuts grown in yearly harvest cycle.
	Minimum 30% rapidly renewable material in Kirei Coco Tiles:
*(MR 7)	**Certified Wood (Pending)**
	Kirei Coco Tiles are manufactured using wood from sustainably managed forests. Kirei is in the process of attaining FSC certification for our entire supply chain. Once this certification is complete our Coco Tile products will be eligible for inclusion in MR 7.

For more information about LEED™, please visit **www.usgbc.org**

Beautiful : Natural : Sustainable   kireiusa.com

- Wood Panel Products, Columbia Forest Products
- Glacier Clear, Plum Creek Timber Company, Inc.
- Green T Particleboard, Timber Products Company
- NU Green, Uniboard
- TemStock-Free Particleboard, Temple-Inland Forest Products
- Terramica, Potlatch Corporation
- Vesta Particleboard, Flakeboard

LEED ONLINE DOCUMENTATION

Once the contractors has documented that all composite wood and agrifiber materials meet the requirements for this LEED credit, they can submit this information by way of the LEED online form for this credit. The first step is to upload all composite wood and agrifiber materials used in the project to the online table. The contractors must certify for each that they contain no added formaldehyde.

The second step involves the contractor certifying that all composite wood and agrifiber materials used in the project have been included in the table. The third step in this process is for the contractors to check off that no laminates used in the project contain added formaldehyde.

GLOSSARY

adhesive   a substance that is able to attach two surfaces together

allergen   any substance that has been known to cause an allergic reaction in humans. These can include various types of mold

anticorrosive coating   a coating formulated to prevent metallic surfaces or substrates from corroding

aquifer   soil or rock formation underground that stores all of the groundwater feeding freshwater springs and wells

architectural coatings   any coating applied to the interior or exterior of a building or component thereof

biochemical oxygen demand   calculates the amount of oxygen living organisms use up in water, and is a measure of pathogenic pollution levels therein

biodegradable material   a material that can be broken-down by a natural decaying process into its original base compounds. Mostly organically based wastes such as food products, paper, and cardboard are biodegradable

biodiversity   the diversity of all forms of life

biofuel-based systems   systems that operate entirely using organic materials, including crops, animal wastes, and gasses from landfills as fuel

biological contaminant   small living organisms including viruses, mold, and bacteria which, when inhaled or ingested, can negatively affect a person's health. These health effects include allergic reactions and respiratory disorders

biomass   a material taken from a plant that can be used to create energy

bituminous coating material   coatings manufactured from crude petroleum oils that are applied to vertical or horizontal surfaces on buildings. These coatings are primarily used for waterproofing

bond breakers   coatings applied to the interior surfaces of concrete formwork to prevent the concrete from bonding to the form. The use of bond breaker coatings allows for the easy removal of concrete forms from hardened concrete without damaging the concrete surface

breathing zone   an area of a room from 3 to 72 in. above the floor and at least 2 ft. from all of the walls or any AC unit

brownfield   an area of land that was either polluted or contaminated with hazardous wastes during its previous use

building automation systems   a system that monitors and maintains building subsystems to make sure that they are operating effectively

building footprint   the area of land that a building covers. This area of land does not include parking lots or landscaping

building indoor air environment   the interior environmental space within a building. These spaces are important to protect from environmental hazards because contaminants introduced into this space can spread rapidly throughout the building and affect many people

bulky wastes   wastes that are too large to be processed by normal municipal solid waste (MSW) methods

carbon dioxide (CO_2) levels   measurements of CO_2 that indicate how effective certain ventilations are. If concentration is at or above 530 parts per million (ppm), the ventilation is inadequate. If the concentration exceeds 800 ppm, then the air quality is poor

carcinogen   any substance that is directly involved in aggravation of cancer. Classified as Group I, 2A, or 2B by the International Agency for research on Cancer

chemical runoff waters that transfer certain chemicals from a project site to local waterways. Some examples of these chemicals include salt and antifreeze

chemical treatment a treatment that manages rusting and scaling by using chemicals

chlorofluorocarbons (CFCs) certain hydrocarbons that are contained in coolants that are harmful to the atmospheric ozone layer

clear wood finishes both totally clear and semitransparent coatings, including varnishes and lacquers. These coating are primarily applied to wood components and surfaces, including cabinetry, wood trim and wood floors

coating a covering that is applied thinly on an object or material to decorate, protect, fill, or seal its surface

co-disposal the disposal of various kinds of waste in the same area of a landfill

collection the process of gathering wastes from both residential and commercial areas, placing them in a truck, and moving them to alternative disposal locations

colorant a concentrated substance that is added to paints or coatings in order to create a specific color

combined heat and power when one fuel source is used to generate both heat and electricity

combustibles materials within piles of waste that can be easily burned or are highly flammable

comfort criteria conditions for human comfort, which include temperature and clothing

commercial adhesive an adhesive substance that is used for any purpose by any manufacturer or business

commingled recycled materials that are gathered together after they were already separated from other municipal solid waste (MSW). These commingled materials are not separated from one another

commingling recycling a system of recycling where people are allowed to place a variety of different materials in the same container for future sorting and recycling

commissioning cycle the organization of commissioning phases during the construction process, which include investigation, analysis, and implementation

common pollutants hazardous gasses and elements, which include CO_2, NO_2, and mercury (Hg)

communal collection a system in which individuals drop off their waste at the central location where it will eventually be collected

compactor vehicles vehicles that are equipped with a high-power waste compacting apparatus that greatly lessens the volume of waste

composite wood wood made from a variety of materials and plant fibers. Some examples include particleboard and plywood

concentrate a product used for cleaning that is diluted before it is actually used

conditioned space an area in a building that is either heated or cooled in order to maintain the comfort of occupants

constructed wetland an artificial system in wetlands that imitates the effects of water treatment and eliminates any polluted waste

construction and demolition (C&D) debris waste that is created by both the construction and demolition of buildings. Examples of such debris include concrete, lumber, wood, and metallic materials

construction IAQ management plan a plan created before a construction project is commenced, in order to reduce or minimize the amount of air contamination in a building

construction, demolition, and land clearing (CDL) debris any waste or material that is generated from construction, demolition, and renovation activities. Any organic matter on-site is also included in the CDL

controlled dumps landfills that are developed in locations that are suitable for the environment, have gas management policies, and control the picking up of wastes

declarant the project team member for LEED who is able to submit a template for credit

development footprint all of the developed area surrounding a building, which includes parking lots, streets, and landscaping

DFT is an acronym for "Dry Film Thickness."

disposal is a process that occurs after collection and processing and is the act of placing solid wastes in a landfill or dump

diversion rate   the fraction of waste that is placed aside for recycling or composting, and of the waste being placed in a landfill or incinerated

drop-off centers   specified areas where wastes that can be recycled or composted are dropped off by those that generate the waste

dump (also see controlled or open dumps)   area of land where solid waste is legally or illegally deposited

durable goods   objects that can be used for long periods of time without being replaced. These objects are usually expensive and include furniture and household appliances

ecological restoration   the process of taking an area of land and restoring it to its original, predevelopment condition

ecologically appropriate site features   the elements found in a site that are a part of nature and are appropriately located

elemental mercury (Hg)   mercury in its purest form, which can be found in fluorescent light bulbs

emissions   gasses that are discharged into Earth's atmosphere

energy recovery   the process of taking energy that can be used elsewhere from waste. A common example is the heat that is extracted from the incineration of wastes

energystar rating   the score given to a building or component, based on its energy performance. Fifty is an average score

environmental impact assessment (EIA)   an assessment that is used to help locate and predict the impact of an action or project on the environment and the health of people

environmental risk assessment (EnRA)   an assessment that evaluates the communications of representatives, people, and resources. It typically evaluates the possibilities of damage that could occur because of the wastes released into the environment

erosion   the process by which solid materials break down into smaller pieces. The term is commonly used to describe the loss or topsoil on a building site

exhaust air   the air that is mechanically removed from a building

existing building commissioning   a process that involves developing a plan that determines the needs of buildings, tests currently used systems, and then provides any necessary changes

facility manager   the individual responsible for supervising multiple building systems

flat coatings   coatings that have a minimum gloss factor. Specifically, flat coatings, when dry, will register a gloss of under 15 at an angle of 85°, or less than 5 at an angle of 60°

floor coatings   any coatings used on a floor surface. These include coatings on interior floors as well as exterior decks

fly ash   the solid waste generated from incinerators. This material is used in the production of concrete

formaldehyde   a volatile organic compound (VOC) that is natural and can cause severe illness in humans

full disclosure   a complete listing of every chemical and its concentration located in a product

fungi   a kingdom of living things that are neither plant nor animal. The kingdom of Fungi includes mushrooms, mold, and yeast

furniture, fixtures, and equipment (FFE)   all objects or items in a building that are movable, but not an actual part of the building. Some examples include desks and chairs

garbage   waste disposed by people, including food, paper, and bulkier items

green cleaning   a process that uses any cleaning product that is safe for the environment and sustainable

groundwater   water that is under the surface of the Earth that fills underground aquifers. Wells and springs both are fed by groundwater

halons   the effective but hazardous chemicals found in some fire suppression systems

hardscape   all of the elements around or in landscaping that are non-plant in nature. Some examples include concrete and brick pavers

hazardous waste   any type of waste that is a danger to people or the environment

heat island effect heat absorption due to the shadow of an object such as a building. The heat increases in areas not within the shadow

heavy metals metals that are dangerous to people or the environment. They contain a high atomic weight and have a greater density. Some examples of these are mercury and cadmium

high-efficiency particulate air (HEPA) filters filters that remove almost 100% of 0.3-μm particles from the air

household hazardous wastes wastes used in residential areas that are harmful to the environment. Some examples are interior paints and toxic cleaning supplies

HVAC systems systems in a building that provide comfort for people. These systems include heating, ventilation, and air-conditioning

HVACR systems systems in a building including heating, ventilation, air-conditioning, and refrigeration

hydrochloroflourocarbons (HCFCs) chemicals in buildings that are used in HVAC systems. They have been found to be hazardous to the atmospheric ozone layer

hydroflourocarbons (HFCs) chemicals used for cooling that are not hazardous to the ozone layer, but do play a role in global warming because of their ability to retain heat in the atmosphere

hydrology the study of the natural occurrence, distribution, and circulation of the water on the Earth and in the atmosphere

hypersensitivity pneumonitis a grouping of diseases that affect the respiratory system causing inflammation of the lung tissue. Most forms of this condition are caused by the inhalation of harmful airborne contaminants, including mold

impervious surfaces a sealed surface that does not allow water or the liquid to penetrate through it

imperviousness the resistance of an object or surface to liquid penetration

in situ remediation use of certain methods to treat contamination on the site

incineration the process of having waste go through a controlled burning in order to lessen its volume. This process is also used to produce energy

incinerator a furnace that is used to burn waste materials

indoor air quality (IAQ) the quality of air in a building. Eighty percent or more of the building's occupants must be satisfied with the quality of air in the building

infiltration the movement of air into or out of a building through openings in the building envelope. It also is a term that refers to water percolating through soil, as in the previous definition

infiltration basins and trenches devices that aid in the removal of storm water from the ground. These devices must clear all water within 72 h.

inorganic wastes wastes that contain matter that is not from plants or animals. Some examples include sand and glass

installation inspections inspections undertaken on buildings to make sure that they are prepared for any performance testing

integrated solid waste management a set of methods used to help manage solid waste

interior stains pigmented coatings that have been designed and manufactured for use on the interior surfaces of buildings

lacquers transparent, protective coatings used on wood that consist of nitrocellulose or resins, which evaporate without going through a chemical reaction

landfill gasses potentially hazardous gasses, for example, methane (CH_4), created when organic wastes decompose

landfilling the last phase of the disposal of waste process, where solid waste is permanently placed in designated area or location

landfills specific sites where waste is disposed of and buried

landscape area the portion of sites designated solely for landscaping, including around parking lots and hardscaped areas

leachate a bacteria or toxic substance-filled and therefore potentially toxic liquid discharge that has leaked from a landfill or pile of waste. It can contaminate any surrounding surface or underground water supply

leed accredited professional (LEED AP)   any individual who has successfully passed the LEED AP examination

Legionella pneumophila   a form of bacteria that grow in slow-moving water and can cause disease

lift   the finished layer of waste that has been compacted in a landfill, but may also refer to a backfilled layer of soil during an earthwork operation in an on-site excavation

liner   a layer of treated soil or synthetic material that is placed all around a landfill in order to protect the groundwater and subsoil environment from the flow of leachate

manual landfills   landfills that are operated almost entirely with equipment that must be worked manually

market waste   any waste that is discarded around food markets. Some examples are fruit leaves and skins and any food that was thrown away because it was not sold

mass-burn incinerator   an incinerator that burns solid waste before it is sorted and processed

mastic coatings   coatings that are used to seal off, or cover small gaps or holes in surfaces. They are applied in a minimum thickness of 10 mils

material safety data sheets (MSDSs)   documents containing warnings associated with certain products, which list all chemicals contained by the product, and includes instructions on how to safely handle, store, and properly dispose of these products

materials recovery   the taking of recoverable materials out of waste, so that they can be recycled and reused later

materials recovery facility (MRF)   a facility that is designed to sort through and sell recycled materials

mechanical ventilation   the process of furnishing air that is produced by mechanical devices like fans

metallic pigmented coatings   coatings that contain a minimum of 48 g/L of metallic, pigmented coating

metering controls   controls that set a limit on how long water can flow

methane   a very flammable, toxic, and highly explosive gas (CH_4) that is produced by decomposing waste at landfills

minimum efficiency reporting value (MERV)   a rating of filtration media which ranges from 1 to 16

mixed waste   a pile of waste that has not yet been sorted

mixed-mode ventilation   a combination of both mechanical and natural ventilation

mold   a group of living organisms that belong to the kingdom of Fungi

monofill   a type of landfill that is designed for only one specific type of waste

MSW   an acronym which stands for "municipal solid waste"

MSWM   an acronym which stands for "municipal solid waste management"

municipal solid waste   consists of any solid waste that was not generated within an industrial or agricultural environment. MSW usually includes debris from construction sites, but does not include wastes that are hazardous and should not enter the waste stream

municipal solid waste management   plans and handles the systems for dealing with waste

MVOC (microbial volatile organic compound)   a chemical gas produced by some types of mold. This gas often has a moldy or musty odor

mycotoxin   a type of toxin produced by mold cells as they multiply. Many mycotoxins have been found to be hazardous to humans if ingested or inhaled

national pollutant discharge elimination system (NPDES)   a program designed to manage the water pollution caused by both industrial and public sources

native and adapted vegetation   plants that are native to a specific area of land, and have adapted to the climate, thus requiring less maintenance

natural ventilation   natural air that circulates in a building without the use of any man-made devices, such as fans

negative pressure   created by mechanical devices in a room to lower the pressure, causing air to flow in, not out

net present value   the total amount of money transferred in and out of a project at a particular point in time

net project material value the total value of specific components of a project, including construction material, furniture, both mechanical and electrical apparatuses, and salvage

nonflat coatings coatings that contain a gloss reading of at least 5 as measured at an angle of 60° and a gloss reading of at least 15 on an angle of 85°

off-gassing the release of volatile organic compounds (VOCs)

off-site salvaged materials objects that are salvaged from one location and used again at another

ongoing commissioning continuous commissioning that allows for optimal building performance

ongoing consumables objects that are inexpensive and are commonly and continuously used in businesses

on-site salvaged materials objects that are salvaged at a particular location, and reused at that same location

open dump an unorganized landfill that usually does not take preventative actions toward leachate and has no system of management, for example, an unattended, vacant tract of land

organic waste waste that is derived from an animal or plant resource and can easily decompose without being too much of a threat to the environment

outdoor air air that travels into buildings from outside

ozone a gas (O_3) created when nitrogen (N) and VOCs react together at ground level. Ozone may also be produced when oxygen (O_2) is exposed to electrical discharges

ozone-depleting substances (ODSs) substances that the United States Environmental Protection Agency (EPA) declares as a possible threat to the atmospheric ozone layer. They contain a depletion potential of at least 0.01

pathogen any organism that has the ability to cause illness or disease

persistent, bioaccumulative, and toxic compounds (PBTs) any substance or compound that is persistent, bioaccumulative, or toxic

perviousness the ability to be pervious, that is, to allow a liquid or gas to penetrate whatever material has perviousness. It also refers to the penetrate through it

plastic cement welding the use of adhesive substances to help dissolve plastic in order to create a bond between two plastic surfaces or objects

plastic foam a foam that was created from plastic materials

pollutants chemicals that are dangerous and hazardous to the environment. Some examples include carbon dioxide (CO_2), sulfur dioxide (SO_2), and mercury (Hg)

porous material a material that contains very small openings, which allow liquids to be absorbed. An example of such a material is wood.

post-consumer content the percentage of material that is created from recycled wastes after the material has been used by the consumer

post-consumer fiber paper and fibrous wastes that were recovered from consumer waste

post-consumer materials materials that consumers have no further use for

potable water water that meets the standards set by the EPA and is suitable for human consumption

PPM acronym that stands for "parts per million," which is the measurement used to gauge the concentration of an element in a liquid or gas substance

previously developed sites sites that were once developed, as opposed to sites in their natural, virgin condition

primary materials marketable materials created from virgin materials in order to produce certain products

primer a substance that improves the bonding capability of a surface

primers coatings that are applied between substrates and subsequent coats in order to provide a solid bond between the two

processing the preparation of municipal solid wastes for further use through crushing, recycling, and other methods

putrescible the tendency for an item or object to decay or decompose quickly

rainforest alliance is an alliance of forestry company whose members agree to protect the environment by planting trees, combating erosion, prohibiting or limiting chemical use, and protecting plants native to the region

rapidly renewable materials an agricultural product that takes a maximum of 10 y. to fully grow and be ready for harvesting

recommissioning the process that occurs when buildings that were previously commissioned need to undergo construction or renovation and be commissioned again

recovered fiber any fiber once used by consumers and industries that can be used again. This includes any used paper products recovered from waste

recyclables items that can be separated from waste and reprocessed to produce a new, reusable product

recycled coatings coatings manufactured by a certified recycled paint company, whose total weight contains at least 50% reused, post-consumer coatings

recycled content the amount of material that was recycled in a product

recycling (1) the process of collecting wastes and sorting out any materials that can be reprocessed and reused later

recycling (2) the reprocessing of certain materials that were separated from waste in order to produce new, reusable products

refrigerants chemical fluids used in devices such as air conditioning and refrigeration equipment

refurbished materials older, already-used products that have been provided with any necessary updating or repairing to increase their lifespan

refuse a term usually used to represent, or mean, "solid waste"

regionally harvested and processed materials any items that are harvested or processed within a maximum of 500 miles from the project site

regionally manufactured products items that are manufactured into final products within a maximum of 500 miles from the project site

remanufactured materials items that are recycled and are used to create new and different products

replacement value the amount it will cost to replace an old product

reproductive toxins a substance, chemical, or agent that can create alterations in the reproductive system of living things

resource conservation and recovery act (RCRA) the legislation that gives the EPA total control over hazardous waste from "cradle to grave"

resource recovery when energy or materials are extracted from wastes

retained components different parts of a building that are reused to help create a new design

retention ponds ponds specifically created to store or collect rain and snow for future use. They are also designed to allow polluted wastes to settle to the bottom (settling basins) for retention and subsequent treatment and disposition

retrofit any updates and replacements executed within a building

return air air circulating indoors that is returned back to the air handling unit

reuse a term that describes when a product in its original form is used more than once, not necessarily for the same purpose

rubbish a term that is used interchangeably with solid waste

rust preventative coatings coatings that are used to prevent the decomposition (corrosion) of metallic surfaces in both residential and commercial settings

sanding sealers clear coatings that are applied directly on wood in order to seal the wood in preparation for any succeeding coatings

sealant a bonding agent that fills and seals gaps between two surfaces, usually to prevent moisture intrusion in buildings

sealers coatings used to prevent materials from breaking through a substrate, as well as used to prevent damage to subsequent coatings

secondary materials materials that are taken from waste in order to replace a prime material, which will aid in the process of creating a new product

secure landfill a type of landfill that makes sure the environment is permanently safe from any potential threats caused by wastes

sedimentation the process by which soil, pebbles, or rock particles are added to, transported by, and deposited by water

sequence of operations an ordering of events described in a building system document that states all of the procedures and variables in the operation of that system

set-out container a containment vessel that is used within residential neighborhoods to collect wastes

set points the term given for a range of values that falls within the norm

shellacs coatings that do not need a chemical reaction in order to dry. They provide firm, protective coatings for sealing out any stains and odors, as well as being used for finishing wood

sick building syndrome an illness that workers obtain in buildings that may contain indoor pollutants or poor ventilation. Some symptoms of this syndrome include headaches and respiratory problems

simple payback the total amount of time taken for an initial investment to be fully recovered though savings

site area the total area of a project, including both built and un-built sections

site energy the amount of a building's heat and electricity that is purchased

soft costs construction costs that are not direct, in that they do not involve materials or labor required by the project. Some examples include architectural, engineering, and permit fees

source reduction (1) minimizes all of the unneeded material in a building

source reduction (2) the lessening of created toxic waste through the recovery and reuse of materials

source separation the sorting of materials from the waste stream by placing those that can be composted and recycled into separate containers

special wastes specific wastes that are not supposed to enter the municipal solid waste stream

spore a primitive, microscopic, usually unicellular, particle molds use in the reproductive process. These particles are released into the air, where they can be easily inhaled or ingested by humans

stains discolorations in materials that do not affect the material's texture

standard operating procedures (SOPs) instructions for any normal operation

stormwater runoff rain or melted snow that does not stay within a project area. It completely leaves the boundary and is not absorbed into soil

sustainable forestry forestry that meets human, animal, and plant needs

sustainable purchasing policies policies that require working with products that are not hazardous and do not damage the environment. Preference is usually given to companies that share this attitude

sustainable purchasing programs methods of purchasing products that are not hazardous and do not damage the environment

systems narrative a verbal or written description of all of the major systems within a building. Some systems examples include heating, cooling, and lighting systems

total phosphorous the quantity of any type of phosphate that is suspended in the water from storms

toxigenic the capability of an organism to produce toxic substances

traffic coatings coatings used on public surfaces including, communal roads, major streets, curbs, pedestrian walkways, and parking lots

transfer station the location where collection trucks drop off MSW to be consolidated. The waste is then separated into loads and transferred to more permanent landfills

transfer stations facilities that obtain, store, then transfer loads of solid wastes and recyclable materials to their appropriate final destinations

two-year, 24-hour design storm the heaviest amount of precipitation that can be statistically expected during a 24-hour period within a 2-year time frame

urea formaldehyde a chemical compound that is commonly used in the manufacturing of particle board and some types of foam insulation, which after incorporation into a building can release large quantities of formaldehyde gas into the building's indoor environment

USDA acronym which stand for: "United States Department of Agriculture"

USDA organic a certification that verifies that certain products are made from a minimum of 95% organic ingredients

varnishes protective wood finishes developed with a variety of resins that, when dried, leave a clear, hard coating

ventilation how air moves within or through a building

virgin materials previously unused, natural materials that are used for industrial processes. Iron ore is an example

volatile organic compounds (VOCs) [1] chemical compounds that have a vapor pressure high enough to completely vaporize and enter the Earth's atmosphere

volatile organic compounds (VOCs) [2] chemicals which, when placed at room temperature, completely evaporate

waste bans declared by the Solid Waste Facility Regulation; it is the prohibition of the discarding of certain materials, including items that can be recycled

waste collector an individual or company who is responsible for collecting the waste from residential neighborhoods, commercial businesses, and community bins

waste dealer an individual or company who buys materials that can be recycled directly from waste generators, and sells them to brokers or recycling plants

waste disposal removing of waste by any method, excluding recycling or reusing

waste diversion placing waste in landfills or incinerators after it has been removed from the waste stream

waste reduction includes all possible means of lessening the amount of municipal solid waste produced

waste reduction program a program designed to reduce the amount of waste diverted to landfills and incinerators. It includes steps that can be taken to lessen waste flow and increase the recycling and reusing of materials

waste stream (2) the amount of total waste that flows through a specific area

waste-to-energy (WTE) plant a facility that generates energy from solid waste. Some WTE plants are able to convert the gas generated from landfills into electricity, whereas others use incinerators to produce the steam used by industries

water table the level beneath the surface of Earth where the ground becomes saturated with water

waterproofing concrete/masonry sealers transparent sealers that are used to seal concrete and stonework in order to make them resistant to water, light, stains, and acids

wetland an area that is almost always wet or saturated with water. It has a water table that is either level with or above the surface of land. Wetlands must also contain certain, specific kinds of vegetation, as defined by the EPA

yard waste organic matter from yards and lawns that has been thrown away. Some examples include cut grass, branches, and leaves

zinc-rich industrial maintenance primers primers whose total weight contains at least 65% metallic zinc dust

BIBLIOGRAPHY

Introduction

1. Landes, Lynn. *River of Waste: A Zero Waste Plan for the Future*. Farmington Hills, MI: Greenhaven Press, 1997.

2. *Municipal Solid Waste in the United States: 2007 Facts and Figures*. Office of Solid Waste, U.S. Environmental Protection Agency, October 2007. http://www.epa.gov/epawaste/nonhaz/municipal/msw99.htm.

3. Ibid.

4. *Landfill Capacity in North America, 1991 Update*. National Solid Waste Management Association, 1991.

5. United States Green Building Council, 2010. USGBC.org.

Chapter 1

1. *2003 Annual Natural Resources Inventory*. United States Department of Agriculture Natural Resources Conservation Service, 2003.

2. *Our Built and Natural Environments*. United States Environmental Protection Agency, 2000.

3. *Environmental Building News*, vol. 5, no. 1 (Updated with 1999 Census data).

4. *2003 Commercial Buildings Energy Consumption Survey—Overview of Commercial Buildings Characteristics*. Energy Information Administration. http://www.eia.doe.gov/emeu/cbecs2003/introduction.html.

5. *C-Series Reports*. Manufacturing and Construction Division, Census Bureau, U.S. Department of Commerce, 1995.

6. *American Housing Survey for the United States—2007*. U.S. Department of Housing and Urban Development and U.S. Department of Commerce, September 2008.

7. U.S. Bureau of the Census.

8. U.S. Forest Stewardship Council.

9. NAHB Research Center's Annual Builder Survey.

10. *Forest Products Journal*. National Forest Products Association, January 2001.

11. *Municipal Solid Waste in the United States: 2007 Facts and Figures*. Office of Solid Waste, U.S. Environmental Protection Agency. October 2007. http://www.epa.gov/epawaste/nonhaz/municipal/msw99.htm.

12. Ibid.

13. Ibid.

14. Ibid.

15. *Buildings Energy Databook, 2006. US Department of Energy and Annual Energy Review 2007*. DOE/EIA-0384 (2007). Energy Information Administration, U.S. Department of Energy, June 2008. http://www.eia.doe.gov/aer/pdf/aer.pdf.

16. Jenkins, P. L., T. J. Phillips, E. J. Mulberg, and S. P. Hui. "Activity Patterns of Californians: Use of and Proximity to Indoor Pollutant Sources." *Atmospheric Environment* 26A (1992): 2141–48.

17. Blasing, T. J. *Recent Greenhouse Gas Concentrations*. Carbon Dioxide Information Analysis Center, December 2009. http://cdiac.ornl.gov/pns/current_ghg.html.

18. Ibid.

19. Intergovernmental Panel on Climate Change, IPCC, 2007.

20. *Inventory of U.S. Greenhouse Gas Emissions and Sinks: 1990–2007*, U.S. EPA.

Chapter 2

1. *LEED 2009: Technical Advancements to the LEED Rating System*. USGBC. http://www.usgbc.org.

2. http://www.clu-in.org/conf/tio/lcia_092309/LEED-2009-Weightings-Tool-Overview.pdf.

3. Kats, Gregory. *Greening America's Schools Costs and Benefits*. A Capital E Report, 2006. http://www.usgbc.org.

4. Kats, Gregory. *Green Building Costs and Financial Benefits*. Capital E, 2003.

5. Ibid.

Chapter 3

1. Koskela, L., G. Howell, G. Ballard, and I. Tommelein. "The Foundation of Lean Construction." In *Design and Construction: Building in Value*, edited by R. Best and G. de Valence. Oxford, UK: Butterworth-Heinemann, Elsevier, 2002.

2. *USA Today*, October 22, 2008. http://www.usatoday.com/travel/news.

3. Meenrajan, Senthil, and Jeff House. "Medical Illness During Travel—A Review." *Northeast Florida Medicine* 60, no. 4 (2009): 10.

Chapter 4

1. *Design Build Effectiveness Study*. U.S. Department of Transportation, January 2006. http://www.fhwa.dot.gov/reports/designbuild/designbuild.htm.

2. Rusie, Candice. "Special Concerns with Green Building Projects." *AGC Legal Brief*, vol. V, no. 3 (2009).

3. *Green Building Disputes Prompt Lawsuits*. July 9, 2009. Environmentalleader.com.

4. Rusie, Candice. "Special Concerns with Green Building Projects." *AGC Legal Brief*, vol. V, no. 3 (2009).

5. Carsman, Howard, and Janet Kim Lin. "Failed Certifications, Increased Costs Can Lead to Lawsuits." *Portland Business Journal*, Friday, November 28, 2008.

6. Sclafane, Susanne. *Insurers "Green Up" Gray Coverage Areas*. Construction carriers tackling unique liability risks of green-building projects. January 2010. http://www.property-casualty.com/Issues/2010/January-11-2009/Pages/Insurers-Green-Up-Gray-Coverage-Areas-.aspx.

Chapter 5

1. *EPA Construction General Permit*. http://cfpub.epa.gov/npdes/stormwater/cgp.cfm.

2. U.S. EPA Publication. "Developing Your Stormwater Pollution Prevention Plan: A Guide for Construction Sites." Document No. EPA-833-R-060-04. http://www.epa.gov/.

3. http://www.green-moxie.com.

4. http://www.erosioncontrolproducts.biz.

5. http://www.priceandcompany.com.

6. Broz, Bob, Don Pfost, and Allen Thompson. *Controlling Runoff and Erosion at Urban Construction Sites*. University of Missouri Biological Engineering Department. http://extension.missouri.edu/publications.

7. *IDEQ Storm Water Best Management Practices*. Idaho Department of Environmental Quality, September 2005. http://www.deq.state.id.us/WATER/data_reports/storm_water/catalog.

Chapter 7

1. The National Recycling Coalition. http://www.nrc-recycle.org.

2. Fact Sheet, Can Manufacturers Institute. http://www.steel.org.

Chapter 10

1. Forest Resources of the United States. National Atlas.gov.

2. Causes of Endangerment. EndangeredSpecie.com.

3. Forest Stewardship Council. http://www.fscus.org/about_us/.

Chapter 11

1. *Air and Radiation*. U.S. Environmental Protection Agency. http://www.epa.gov/air.

2. Ibid.

3. *Sick Building Syndrome*. United States Environmental Protection Agency. http://www.epa.gov/iaq/pubs/sbs.html.

4. American Lung Association, State of Lung Disease in Diverse Communities, 2010. http://www.lungusa.org/assets/documents/publications/solddc-chapters/occupational.pdf.

5. *Patient Facts: Learn More about Legionnaires' Disease*. Department of Health and Human Services, Centers for Disease Control and Prevention. http://www.cdc.gov/legionella/patient_facts.htm.

6. *Indoor Air Quality (IAQ) Market Research.* National Energy Management Institute (NEMI), June 2002. www.nemionline.org.
7. U.S. Department of Labor, Occupational Safety and Health Administration. http://www.osha.gov/dts/osta/otm/legionnaires.

Chapter 12
1. University of Washington Department of Environmental and Occupational Health Sciences. http://depts.washington.edu/silica.

Chapter 15
1. Carpet and Rug Institute (CRI). http://www.carpet-rug.org/about-cri/the-history-of-carpet.cfm.

2. Ibid.
3. Ibid.
4. Ibid.
5. SCS-EC10.2-2007, Environmental Certification Program, Indoor Air Quality Performance Scientific Certification Systems (SCS). http://www.scscertified.com/gbc/docs.
6. Carpet and Rug Institute, CRI. http://www.carpet-rug.org.
7. Ibid.
8. Ibid.

INDEX

Note: The letter 'f' following a page number indicates an illustration or a photograph.